Magnetohydrodynamic Equilibrium
and Stability of Stellarators

Frances Bauer
Octavio Betancourt
Paul Garabedian

Magnetohydrodynamic Equilibrium and Stability of Stellarators

With 25 Illustrations

Springer-Verlag
New York Berlin Heidelberg Tokyo

Frances Bauer
Octavio Betancourt
Paul Garabedian
Courant Institute of Mathematical Sciences
New York University
New York, NY 10012
USA

Library of Congress Cataloging in Publication Data

Bauer, Frances.
  Magnetohydrodynamic equilibrium and stability of
stellarators.

  Bibliography: p.
  Includes index.
  1. Stellarators—Mathematical models. 2. Plasma
confinement—Mathematical models. 3. Stellarators—
Data processing. 4. Plasma confinement—Data processing.
I. Betancourt, Octavio, 1945   . II. Garabcdian,
Paul. III. Title.
QC791.77.S7B38   1984   621.48′4   84-1236

© 1984 by Springer-Verlag New York Inc.
Softcover reprint of the hardcover 1st edition 1984

Typeset by Science Typographers, Inc., Medford, New York.

9 8 7 6 5 4 3 2 1

ISBN-13: 978-1-4612-9753-6   e-ISBN-13: 978-1-4612-5240-5
DOI: 10.1007/978-1-4612-5240-5

To Cathy, Emily and Sofia

Preface

In this book, we describe in detail a numerical method to study the equilibrium and stability of a plasma confined by a strong magnetic field in toroidal geometry without two-dimensional symmetry. The principal application is to stellarators, which are currently of interest in thermonuclear fusion research. Our mathematical model is based on the partial differential equations of ideal magnetohydrodynamics. The main contribution is a computer code named BETA that is listed in the final chapter.

This work is the natural continuation of an investigation that was presented in an early volume of the Springer Series in Computational Physics (cf. [3]). It has been supported over a period of years by the U.S. Department of Energy under Contract DE-AC02-76ER03077 with New York University. We would like to express our gratitude to Dr. Franz Herrnegger for the assistance he has given us with the preparation of the manuscript. We are especially indebted to Connie Engle for the high quality of the final typescript.

New York   F. BAUER
October 1983   O. BETANCOURT
                                                     P. GARABEDIAN

Contents

CHAPTER 1
Introduction

Research on controlled thermonuclear fusion has been pursued on a large scale for several decades. A standard approach is through confinement of plasma in a strong magnetic field. For most configurations the plasma is contained inside a family of toroidal magnetic surfaces. Radial outward shift is compensated for either by a net current, as in the tokamak, or by helical coils like those of the classical stellarator. Three-dimensional effects become important in the latter case. It is the purpose of this book to develop a mathematical and computational model of such configurations that is based on ideal magnetohydrodynamics. Central to the theory is the computer code BETA for the calculation of equilibrium and stability without any assumption of two-dimensional symmetry. The material to be presented represents a self-contained exposition of results obtained since the appearance of an earlier volume on this subject [3].

Issues of major interest for plasma confinement are magnetohydrodynamic equilibrium and stability, and transport. In the past they have been investigated by means of expansions in small parameters and analysis in two space dimensions. More recently large computer codes have been accorded wide acceptance as a tool to solve harder problems in three dimensions. Standard codes to trace magnetic lines of force or to follow particle orbits and assess transport by Monte Carlo techniques are now available.

The problem of magnetohydrodynamic equilibrium and stability in fully three-dimensional geometry, to which this book is devoted, has been treated numerically by applying the classical variational principle of magnetohydrodynamics. Our approach is to introduce flux coordinates describing the equilibrium and to estimate stability by either surveying the energy landscape or applying the Mercier criterion [5, 11]. We are able to handle nonlinear stability, and a sharp boundary model including both vacuum and plasma regions is discussed. Another numerical method based on the variational principle has been developed by Chodura and Schlüter [22], but their code is restricted to a plasma region, uses Eulerian coordinates, and solves for the pressure in a significantly different fashion.

In the next two chapters of the book we review the fundamentals of our computational method in magnetohydrodynamics from a new perspective. The two chapters that follow are devoted to a quantitative treatment of nonlinear stability based on the minimum energy principle and to a discussion of the Mercier criterion for local stability. An analysis of nested flux surfaces, magnetic islands, and transport is appended. After that appear three chapters on stellarators. First a general description of major stellarator experiments is presented. Then there is a chapter on mathematical models of the WISTOR-U proposal from the University of Wisconsin and the ATF-1 experiment to be built at the Oak Ridge National Laboratory. This is followed by an investigation of the Heliac designed at the Princeton Plasma Physics Laboratory, which seems to offer a way of achieving stable confinement at high pressure. The final four chapters of the book are allocated to a FORTRAN listing of a new version of our computer code BETA, together with a sample run, a user's manual, and a glossary of parameters. The aim is to present adequate detail about the method, its implementation, and its applications.

CHAPTER 2

Synopsis of the Method

1. Variational Principle

We consider a formulation of the variational principle of magnetohydrodynamics that enables us to calculate equilibrium and stability without recourse to the full equations of motion [3, 37]. Denote the magnetic field by B, the fluid pressure by p, and the mass density by ρ. We assume an equation of state of the form $p = \rho^\gamma$, where γ is the gas constant. Let

$$E_P = \iiint \left[\frac{B^2}{2} + \frac{p}{\gamma - 1} \right] dx\, dy\, dz$$

stand for the potential energy of a torus of plasma and let

$$E_V = \iiint \frac{B^2}{2} dx\, dy\, dz$$

stand for the potential energy of the magnetic field in a vacuum region surrounding the plasma. The normal component of B is supposed to vanish at the free boundary separating the plasma from the vacuum. Subject to constraints on B, p and ρ that will be formulated presently, the variational principle asserts that equilibrium is characterized by a stationary value of the Hamiltonian

$$E = E_P - E_V.$$

Moreover, the equilibrium is stable if and only if E becomes a relative minimum [3, 9].

It is required that

$$\nabla \cdot B = 0$$

in the plasma, so we can introduce Clebsch flux functions s and ψ there such that

$$B = \nabla s \times \nabla \psi.$$

The first function s is supposed to be single-valued and labels magnetic surfaces

$$s = \text{const.}$$

which are supposed to comprise a nested family of tori. We shall have occasion to discuss the significance of this nested surface hypothesis later on. Let u and v be poloidal and toroidal coordinates that have unit periods the short way and the long way around the nested tori. In practice we take v proportional to the angle in a cylindrical coordinate system about the major axis of the plasma. The second flux function ψ is supposed to have periods specified by the formula

$$\psi = -u + \mu v + f,$$

where $f = f(s, u, v)$ is single-valued. The coefficient $\mu = \mu(s)$, which depends on s alone, is called the rotational transform because it is a measure of the amount through which a magnetic line rotates in the poloidal direction when it makes a complete circuit around the plasma in the toroidal direction. The rotational transform is seen to be the derivative of the poloidal flux between two magnetic surfaces $s = \text{const.}$ with respect to the corresponding toroidal flux, which reduces to s itself. In much of the stellarator literature it is denoted by the symbol ι instead of the letter μ.

The partial differential equations of ideal magnetohydrodynamics imply that the mass inside each flux surface $s = \text{const.}$ is conserved, so we must impose a corresponding constraint on the density ρ in our variational principle. For any fixed choice of the magnetic field B, minimization of the potential energy E_p with respect to p separately subject to this constraint shows that p and ρ become functions of s alone. Without significant loss of generality we therefore begin with the ergodic constraint that ρ is a function of s associated with some assigned distribution of mass [3, 11]. This formulation of the problem is suggested by the equilibrium equation $B \cdot \nabla p = 0$. The term ergodic is motivated by the case of lines of force covering a surface with irrational μ.

A similar version of the variational principle which yields more or less equivalent results about equilibrium and stability can be obtained by putting $\gamma = 0$, or in other words by setting

$$E_P = \iiint \left[\frac{B^2}{2} - p \right] dx\, dy\, dz,$$

and then by requiring $p = p(s)$ to be a given function of s. This has the advantage of eliminating explicit reference to mass and density [27]. In the applications special importance is attached to the average plasma parameter beta defined to be

$$\beta = \iiint \frac{2p}{B^2 + 2p} ds\, du\, dv.$$

The remaining constraint to be imposed in the plasma is that the poloidal flux be a prescribed function of the toroidal flux s, which amounts to prescribing the rotational transform $\mu = \mu(s)$. This form of the variational principle is appropriate for tokamaks because they have net toroidal current $I = I(s)$ different from zero. Fixing μ is also appropriate in any consideration of stability. However, for stellarators it is customary to require

$$I(s) = 0.$$

That requirement can be achieved by an additional minimization of E_P with respect to μ. We therefore allow for such an option in calculations of stellarator equilibrium.

The normalization of the potential energy E_V of the field in the vacuum surrounding the plasma is somewhat different. We put

$$B = \nabla \phi$$

there and solve Laplace's equation

$$\Delta \phi = 0$$

for the scalar potential ϕ. A Dirichlet boundary condition is imposed on ϕ at the outer wall of a toroidal shell comprising the vacuum region to model various winding laws that specify coils maintaining the magnetic field. Also, the normal derivative of ϕ must vanish at the free boundary between the vacuum and the plasma. The potential ϕ is supposed to have fixed poloidal and toroidal periods, which explains why E_V appears with a minus sign in the definition of the Hamiltonian E. For stellarators the poloidal period of ϕ should be zero because it represents a net toroidal current

$$I = \oint d\phi.$$

An application of the calculus of variations shows that E_P becomes stationary when the equilibrium equations

$$J \times B = \nabla p$$

of magnetostatics are fulfilled inside the plasma, where

$$J = \nabla \times B$$

is the current density. The result remains true in the important special case where the outer surface of the plasma is held fixed. On the other hand, if the surface between the plasma and the vacuum is allowed to move freely, the variational principle yields the free boundary condition

$$\frac{1}{2} B_P^2 + p = \frac{1}{2} B_V^2$$

relating the field B_P in the plasma to the field B_V in the vacuum at their interface. This follows from the variational formulas

$$\delta E_P = \iint \left[\frac{1}{2} B_P^2 + p \right] \delta \nu \, d\sigma,$$

$$\delta E_V = \iint \frac{1}{2} B_V^2 \delta \nu \, d\sigma$$

for E_P and E_V under an inward normal shift $\delta\nu$ of the interface, where $d\sigma$ stands for the surface element. As indicated above, we have

$$\nabla \times B = \nabla \cdot B = 0$$

inside the vacuum.

The form of the magnetostatic equations in the plasma shows that the pressure p is a flux function, which helps to justify our original assumption that a single-valued choice could be found for the Clebsch function s. Notice that these equations yield the ordinary differential equation

$$B \cdot \nabla p = 0$$

along the magnetic field lines, which asserts that they are real characteristics of the full system. Our ergodic constraint $p = p(s)$ serves to solve the ordinary differential equation for p and thus eliminates the real characteristics. This is desirable from the point of view of formulating a well posed boundary value problem to determine an equilibrium solution. The pair of partial differential equations we obtain for the two unknowns s and ψ is still of somewhat nonstandard type, but it is more tractable numerically than the system of eight equations in eight unknowns usually considered in magnetohydrodynamics. Further questions about existence of solutions of the equations for s and ψ, which must satisfy the nested surface hypothesis we introduced above, involve a deeper analysis of the KAM theory of dynamical systems [16, 28, 41]. Suffice it to say that our numerical method produces only an approximate solution which may deteriorate as the computational mesh is refined (cf. Sections 3.1 and 6.2).

2. Coordinate System

Consider a modified system of cylindrical coordinates r, θ and z defined by the formulas

$$x = (A + r)\cos\theta, \qquad y = (A + r)\sin\theta, \qquad z = z.$$

The origin has been transformed to a circle of radius A around the z-axis, and z is measured in the direction of the major axis of a torus whose major radius is A. Put

$$v = \frac{Q\theta}{2\pi}$$

for some number of field periods $Q > 0$, let

$$r + iz = r_0(v) + iz_0(v)$$

describe the magnetic axis $s = 0$, and let

$$r + iz = r_1(u, v) + iz_1(u, v)$$

describe the plasma surface $s = 1$. We introduce a function $R = R(s, u, v)$ in the range $0 \le R \le 1$ such that the formula

$$r + iz = r_0 + iz_0 + R[r_1 + iz_1 - r_0 - iz_0]$$

defines a parametric representation of the flux surfaces $s =$ const., with

$$R(0, u, v) = 0, \qquad R(1, u, v) = 1.$$

This representation presupposes a starlike property of the surfaces relative to the magnetic axis that supersedes our nested surface hypothesis [14].

It is useful to include the straight periodic cylinder as a limiting case of the toroidal configuration. Let $\varepsilon = 1/\Lambda$ be the inverse aspect ratio so $2\pi\Lambda/Q$ is the length of a field period. The geometry may then be expressed in terms of ε, with $\varepsilon = 0$ representing the straight case.

Another version of the code that provides better zoning is obtained by putting

$$r + iz = \left(r_0 + iz_0 + Re^{2\pi iu}\right)e^{2\pi niv}$$

with R evaluated at nodes and ψ evaluated at centers of mesh intervals in s. This allows for the two-dimensional calculation of helically symmetric equilibria. In both formulations a major difficulty arises in the treatment of the magnetic axis. An argument from the calculus of variations shows that the condition for equilibrium at the magnetic axis reduces to a requirement of the form

$$\int \left(\frac{1}{2}B^2 + p\right)(dr + i\,dz) = 0,$$

where the line integral is taken over an infinitesimal circle surrounding the point $r = r_0$, $z = z_0$ in each plane section $v =$ const. Related issues concerning the behavior of R at $s = 0$ will be discussed in Section 3.1.

Similar coordinates may be introduced in the vacuum region surrounding the plasma by setting

$$r + iz = r_1 + iz_1 + s(r_2 + iz_2 - r_1 - iz_1), \qquad 0 \le s \le 1,$$

there with r_2 and z_2 taken as parametric functions specifying the shape of an outer control surface. The free boundary of the plasma can in turn be represented in the form

$$r_1 + iz_1 = r_3 + iz_3 + g(r_2 + iz_2 - r_3 - iz_3),$$

where $r_3 = r_3(v)$ and $z_3 = z_3(v)$ are the equations of a curve we shall call the vacuum axis and $g = g(u, v)$ is a function controlling the shape of the free boundary.

The Hamiltonian E becomes a functional of all of the new unknowns R, ψ, g, r_0 and z_0 over a computational cube

$$0 \le s \le 1, \qquad 0 \le u \le 1, \qquad 0 \le v \le 1$$

in the domain of the flux coordinates s, u and v. In the next chapter we shall explain how equations determining these unknowns can be derived from the variational principle of magnetohydrodynamics in a fashion that lends itself to numerical calculation.

CHAPTER 3
Finite Difference Scheme

1. Difference Equations

To arrive at finite difference equations modeling magnetohydrodynamic equilibrium we use a technique that is motivated by the finite element method [3]. First we develop a second order accurate numerical quadrature formula for the Hamiltonian E based on a rectangular grid of mesh points over a unit cube of the space with coordinates s, u and v. We differentiate that formula with respect to nodal values of the unknowns R, ψ, r_0 and z_0 in order to derive difference approximations to the magnetostatic equations. This procedure yields equations in a conservation form that is automatically compatible with conditions stemming from the fact that the flux function ψ is only determined up to an arbitrary additive function of s. Thus we avoid elementary difficulties with existence of solutions of the equilibrium problem which tend to obscure the more serious issues raised by the KAM theory.

   The numerical method is sufficiently successful in practice to suggest that an asymptotically valid solution can be found which may be physically meaningful for small though nonzero values of the mesh size h. Truncation errors are to be reduced to a level comparable to the physical effects of resistivity and finite gyroradius. In this connection observe that our formulation of the flux constraints of ideal magnetohydrodynamics eliminates numerical resistivity from the model. On the other hand, the method produces substantial artificial viscosity that stabilizes equilibria to a degree that is quite perceptible for most configurations of practical interest. Thus it is necessary to perform elaborate convergence studies before making any final conclusion about questions of stability.

   We have chosen to express B in the contravariant form

$$B = \nabla s \times \nabla \psi = \psi_u \nabla s \times \nabla u + \psi_v \nabla s \times \nabla v,$$

but it also has a covariant representation

$$B = B_1 \nabla s + B_2 \nabla u + B_3 \nabla v.$$

In this notation we can write

$$E_P = \iiint \left[\frac{B_3 \psi_u - B_2 \psi_v}{2} + \frac{pD}{\gamma - 1} \right] ds \, du \, dv$$

$$= \iiint \left[\frac{A_{11} \psi_u^2 + 2 A_{12} \psi_u \psi_v + A_{22} \psi_v^2}{2D} + \frac{pD}{\gamma - 1} \right] ds \, du \, dv,$$

where the coefficients

$$A_{11} = r_v^2 + z_v^2 + (A + r)^2 \theta_v^2,$$

$$A_{12} = -r_u r_v - z_u z_v,$$

$$A_{22} = r_u^2 + z_u^2$$

depend on R and $D = \partial(x, y, z)/\partial(s, u, v)$ is the Jacobian whose reciprocal is given by

$$\frac{1}{D} = \nabla s \cdot (\nabla u \times \nabla v) = \frac{G}{RR_s}$$

with

$$\frac{1}{G} = \left[(r_1 - r_0) \frac{\partial z_1}{\partial u} - (z_1 - z_0) \frac{\partial r_1}{\partial u} \right] (A + r) \theta_v.$$

The contravariant representation of the current density is

$$J = \nabla B_1 \times \nabla s + \nabla B_2 \times \nabla v + \nabla B_3 \times \nabla v$$

$$= \left(\frac{\partial B_2}{\partial s} - \frac{\partial B_1}{\partial u} \right) \nabla s \times \nabla u + \left(\frac{\partial B_3}{\partial s} - \frac{\partial B_1}{\partial v} \right) \nabla s \times \nabla v$$

$$+ \left(\frac{\partial B_3}{\partial u} - \frac{\partial B_2}{\partial v} \right) \nabla u \times \nabla v.$$

Our discrete analogue of the energy E_P is based on a mixture of theory and numerical experimentation. For details we refer to the listing of the code in Chapter 14. Suffice it to say that the primary difficulty occurs at the magnetic axis $s = 0$, which becomes a singularity in our computational coordinate system. That difficulty is resolved in the code by putting $R^2 = sW$ and introducing a finite difference approximation of the form

$$\frac{G}{RR_s} = \frac{(s_2 - s_1) G_1}{(s_2 - s_1) W_1 + s_1 (W_2 - W_1)} + \frac{(s_2 - s_1) G_2}{(s_2 - s_1) W_2 + s_2 (W_2 - W_1)}$$

for the important term $1/D$, where the subscripts indicate nodal values. Moreover, at $s = 0$ we use a quadratic expression

$$\frac{1}{W} = c_{11} (r_1 - r_0)^2 + 2 c_{12} (r_1 - r_0)(z_1 - z_0) + c_{22} (z_1 - z_0)^2$$

for the reciprocal of W that is suggested by the Taylor series expansion of s as a function of θ, r and z. The scheme has been devised to assure convergence of the magnetic axis.

2. Island Structure

Our assumption that for each v the toroidal magnetic surfaces are starlike and can be represented nonparametrically by a function R of the flux coordinate s may be formulated analytically as a one-sided constraint $R_s \geq 0$ in the variational principle. Numerically this corresponds to the observation that the two denominators in our second order accurate finite difference approximation to G/RR_s both tend to remain positive. Since each of these denominators separately provides only first order accuracy, it turns out that the method we have described smears out magnetic islands substantially. The islands consist of smaller tori separated from the rest of the field by a large flux surface that intersects itself. They violate the nested surface hypothesis, but they do have the tendency to form in the neighborhood of resonant surfaces where the rotational transform μ takes on rational values n/m associated with small m.

Better resolution of the islands can be achieved by evaluating ψ at centers rather than at nodes of the mesh intervals in s and by introducing the simpler approximation

$$\frac{G}{RR_s} = \frac{(s_2 - s_1)(G_1 + G_2)}{s_2 W_2 - s_1 W_1}$$

of the Jacobian that has a single second order accurate denominator. Results based on this formula, which unfortunately leads to poor convergence of the magnetic axis, are presented in Fig. 1. Shown there are four cross sections of the flux surfaces of an equilibrium that is bounded by a helically perturbed circular cylinder. The equilibrium has an $m = n = 1$ island that appears as a twisted crescent bounded by two magnetic surfaces which merge on the opposite side of the cylinder to form a current sheet. The Jacobian D is large across the crescent, but becomes small at the current sheet. The scheme provides an effective method of capturing current sheets or vortex sheets numerically. For this purpose it may be preferable to spectral techniques that are otherwise advantageous.

In the case of zero pressure the size of any magnetic islands that may occur in a computed equilibrium can be estimated by Fourier analysis even when the original difference scheme we have described is used. It follows from the magnetostatic equations that

$$B \cdot \nabla \frac{J \cdot B}{B^2} = (\nabla p \times B) \cdot \nabla \frac{1}{B^2}.$$

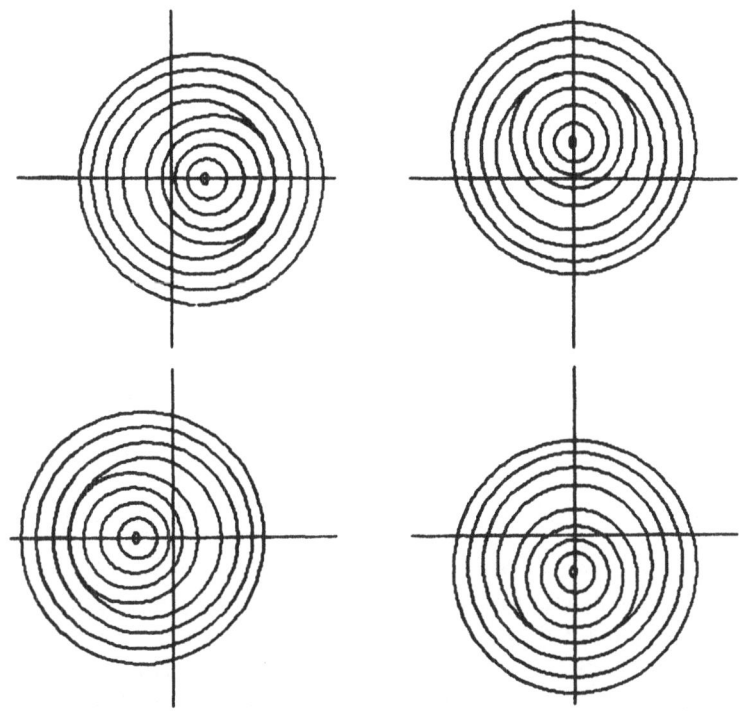

CROSS SECTIONS AT V= .00,.25,.50,.75, QLZ/2*PI= 0.50
MAJOR RADIUS INFINITE   MINOR RADIUS= 1.00

Figure 1. Resolution of an $m = n = 1$ magnetic island.

This means that $J \cdot B / B^2$ must remain constant on any toroidal flux surface where $\nabla p = 0$ and where μ is irrational, since the magnetic lines of force on such a surface are ergodic. Thus if there are no islands we should expect that $J \cdot B / B^2$ is a function of s alone. Consequently the norm

$$d = \left[\frac{1}{16} \iiint \left| \frac{J \cdot B}{B^2} - \iint \frac{J \cdot B}{B^2} \, du \, dv \right|^2 ds \, du \, dv \right]^{1/2}$$

ought to vanish. Conversely, if $d \neq 0$ we conclude that islands must be present.

By examining special cases we have found empirically that the standard deviation d is a reasonable measure of the total width of the islands appearing in the plasma [12]. Roughly speaking, any equilibrium with $d < 0.1$ seems to be fairly well approximated by calculations based on the nested surface hypothesis. On the other hand, when $d > 0.3$ the islands can be expected to overlap so that disruption may take place. Dimensional analysis suggests that we might interpret d^2 as a sort of anomalous diffusion coefficient describing transport phenomena associated with islands and ergodic regions. In any case, it becomes apparent that diagnostics are available to test for the presence of islands in our equilibrium calculations despite the nested surface hypothesis that has been made (cf. Sections 5.3 and 6.2).

3. Accelerated Iteration Procedure

Difference equations for the magnetic field B in the vacuum region are derived by differentiating the Hamiltonian E with respect to nodal values of the scalar potential ϕ. This has the advantage of yielding a compatible system even when only Neumann boundary conditions are imposed on ϕ. However, the free boundary condition

$$M[g, g_u, g_v] = \frac{1}{2}B_P^2 + p - \frac{1}{2}B_V^2 = 0$$

is handled independently of the finite element method because the formulas for derivatives of E with respect to nodal values of the free surface function g are cumbersome. Consequently it turns out in practice that the finite difference equations for fixed boundary equilibria involving only the unknowns R, ψ, r_0 and z_0 can be solved to any desired degree of accuracy, whereas the difference equations for free boundary equilibria incorporate incompatibilities that limit the level to which their residuals can be reduced at any specified mesh size. The incompatibilities are related to incorrectness of boundary value problems for single-valued solutions of a first order partial differential equation on a torus.

An iterative procedure to solve the finite difference equations characterizing equilibrium has been developed by applying the method of steepest descent, or, rather, the conjugate gradient method, to the variational principle of magnetohydrodynamics [3]. The scheme is best described by a continuous model involving an artificial time parameter t that indicates the number of iterations that have been performed. It is similar to the second order Richardson method.

Let U stand for the vector of unknowns R, ψ, r_0 and z_0 and let

$$L[U] = 0$$

stand for the magnetostatic equations in the plasma. Our iterative scheme is a finite difference analogue of the system of partial differential equations

$$aU_{tt} + eU_t = L[U].$$

The coefficient a is chosen to make the system of the hyperbolic type, and the coefficient e is chosen to achieve descent of E_P to a steady state. In practice we have implemented an acceleration procedure that optimizes the choice of e as a function of t.

Let us consider a single linear equation with scalar coefficients a and e. In the case of a system of equations, a and e are diagonal matrices with the elements of e proportional to those of a. As described in [3], the choice of e that maximizes the convergence rate is related to the lowest eigenvalue $-\tilde{\omega}_0^2$ of the operator L by $e/a = 2|\tilde{\omega}_0|$. However, we also need e/a large enough to assure descent in the initial nonlinear phase and to eliminate oscillations due to large eigenvalues. Therefore, we want to choose e/a proportional to the dominant eigenvalue $|\tilde{\omega}|$ at any time in the iteration. To estimate $|\tilde{\omega}|$ we consider

$$F(t) = \int U_t^2 \, dx$$

and the approximation at time step $t = t_j$

$$-\tilde{\omega}_j^2 = (F_t/2F + e/a)_t + (F_t/2F)(F_t/2F + e/a).$$

Finally, we choose e/a at $t = t_j$ proportional to the average of $|\tilde{\omega}_j^2|$ over the previous N time steps,

$$\frac{e}{a} = 2\left(\frac{1}{N} \sum_k |\tilde{\omega}_{j-k}^2| \right)^{1/2}.$$

The initial values of e/a are chosen to be large and constant in time through the initial nonlinear phase of the iteration. The acceleration scheme is then activated, with the large frequencies decaying rapidly. The bulk of the computation consists of the slower convergence of the eigenfunction associated with the lowest eigenvalue of L, with e/a practically constant and between one and two orders of magnitude smaller than its initial values. The number of iterative cycles needed to achieve a desired degree of accuracy in the solution of the difference equations becomes proportional to the reciprocal of the mesh size h, and the amount of computer time required to calculate fully three-dimensional equilibria scales like h^{-4}.

An application of the calculus of variations shows that the change in the plasma energy due to a perturbation $\delta\mu$ of the rotational transform is given by the formula

$$\delta E_P = \int I \, \delta\mu \, ds.$$

Therefore we introduce a discrete approximation to the equation

$$\mu_t = -I$$

as an iterative scheme to achieve the stellarator condition $I(s) = 0$ [4].

In the vacuum region we use successive overrelaxation to solve Laplace's equation for ϕ. Since the vacuum energy E_V is a minimum, the Hamiltonian $E = E_P - E_V$ is a maximum with respect to perturbations of the vacuum potential. It follows that free boundary equilibria correspond to saddle points of the Hamiltonian E. This means that Laplace's equation must be solved accurately for each position of the free boundary. In practice it usually suffices to do three iterations on Laplace's equation for each iteration of the free boundary.

The free boundary is determined by an iterative procedure modeling the partial differential equation

$$g_t = M + b\left(M_{g_u} M_u + M_{g_v} M_v \right)$$

for g. An analogue of the Lax–Wendroff finite difference scheme is applied to calculate g, with the parameter b selected to achieve adequate convergence. A minimum of artificial surface tension is introduced. Convergence of the numerical calculations gives some information about the existence of solutions defining magnetohydrodynamic equilibria. However, questions of stability are hard to answer by tracking the dependence of the iterative solution on the artificial time parameter t because truncation errors are significant in most cases of physical interest. In the next chapter we shall describe how this difficulty can be overcome by a further appeal to the variational principle.

CHAPTER 4
Nonlinear Stability

1. Second Minimization

For meshes that fit on present computers there are appreciable truncation errors in the method we have developed to calculate magnetohydrodynamic equilibrium in three dimensions. These errors comprise an artificial viscosity that tends to stabilize the equilibria numerically. Therefore it becomes necessary to perform careful convergence studies before drawing any conclusions about stability. To that end it is desirable to introduce a quantitative measure of the stability of equilibria on any specific mesh so that extrapolation to zero mesh size becomes feasible. No adequate assessment of growth rates can be made by examination of the dependence of the iterative solution of the problem on artificial time. A more successful procedure is to estimate growth rates by means of a form of the Rayleigh quotient that follows in a natural way from the variational principle of magnetohydrodynamics [4]. We shall describe such a procedure in this chapter and we shall discuss its application to questions of both linear and nonlinear stability.

Let E_0 be the extremal value of the potential energy E_P of an equilibrium whose stability is under investigation. With the surface of the plasma held fixed, let ξ stand for a perturbation δR, $\delta\psi$, δr_0 and δz_0 of the quantities R, ψ, r_0 and z_0 characterizing the equilibrium. Denote by ξ_0 some test function corresponding to an internal mode of perturbation of the equilibrium that may be unstable. In linear stability theory an analogous displacement $\tilde{\xi}$ is given by the formula

$$\delta B = \nabla \times (\tilde{\xi} \times B), \qquad \tilde{\xi} \times B = \delta s\, \nabla\psi - \delta\psi\, \nabla s.$$

Motivated by this consideration, we introduce the norm [5]

$$\|\xi\|^2 = \iiint \left[(\delta R)^2 + w_1(s)(\delta\psi)^2 \right] \rho D \, ds \, du \, dv + w_2 \int \left[(\delta r_0)^2 + (\delta z_0)^2 \right] dv,$$

where the weight factors $w_1(s)$, w_2 have been chosen in a fashion suggested by the case of a screw pinch. This simple choice of the norm leads to a Rayleigh quotient that provides a reasonably good approximation of actual eigenvalues [13].

Let (ξ, ξ_0) stand for the corresponding scalar product. Given $\varepsilon_0 > 0$, we perform a second minimization of E_P with respect to ξ subject to the linear constraint

$$(\xi, \xi_0) = \varepsilon_0.$$

The solution defines a perturbed value E_1 of the energy. The sign of the variation in energy $\delta W = E_1 - E_0$ has a bearing on the nonlinear stability of the equilibrium and the generalized Rayleigh quotient

$$-\omega^2 = 2 \frac{E_1 - E_0}{\|\xi\|^2}$$

specifies a growth rate ω associated with the test function ξ_0 and the amplitude ε_0. This constrained extremal problem for ξ can be solved by the same iterative scheme that was used to calculate the equilibrium, except that a projection must be included to keep the iterates within the hyperplane of the constraint. The solution ξ satisfies an equation of the form

$$L(\xi) = w\xi_0.$$

For small ε_0, after division by the Alfven speed, the quantity ω becomes well correlated with the true linear growth rate. We can also explore nonlinear behavior by fixing ξ_0 and studying the dependence of $-\omega^2$ on the amplitude of the perturbation ε_0.

The procedure we have proposed in order to study stability becomes foolproof when it is applied to determine whether a quadratic form

$$E = \sum \lambda_j a_j^2$$

in the variables $a_1, \ldots, a_N$ is positive-definite. In fact, let us minimize E subject to a linear constraint of the form

$$\sum b_j a_j = \varepsilon_0.$$

If some λ_j is negative, as in the case of instability, then E can be made negative by appropriate selection of the corresponding coefficient a_j without violating the constraint. The result holds regardless of the choice of the test vector $(b_1, \ldots, b_N)$, for $E \to -\infty$ if $b_j = 0$. On the other hand, when all the λ_j are positive the method of Lagrange multipliers shows that the extremal value E_1 of E is

$$E_1 = \frac{\varepsilon_0^2}{\sum b_j^2/\lambda_j}.$$

Notice that a good approximation to the smallest eigenvalue λ_j may be obtained by forming a Rayleigh quotient from this expression, which becomes stationary when $(b_1,\ldots,b_N)$ reduces to the corresponding eigenvector.

2. Test Functions and Convergence Studies

The situation becomes more complicated in the general case, especially if there is nonlinear saturation of the unstable modes. Theorems about numerical optimization show that it is essential to use a linear constraint in formulating the auxiliary extremal problem for E_1. Our procedure is to minimize E_P over all perturbations ξ lying in some hyperplane of function space. In practice this second minimization may fail to reveal any region $E_1 < E_0$ indicating instability if the amplitude ε_0 is too large and the test function ξ_0 differs appreciably from any unstable mode (cf. Fig. 2). The difficulty can be partially overcome by repeating the minimization for several different choices of the test function ξ_0, so that the search for negative values of $\delta W = E_1 - E_0$ is conducted over a larger set surrounding the equilibrium. However, the issue is further obscured in practice by truncation errors associated with our numerical scheme. Thus it becomes necessary to study the dependence of the growth rate ω on the mesh size h and to extrapolate.

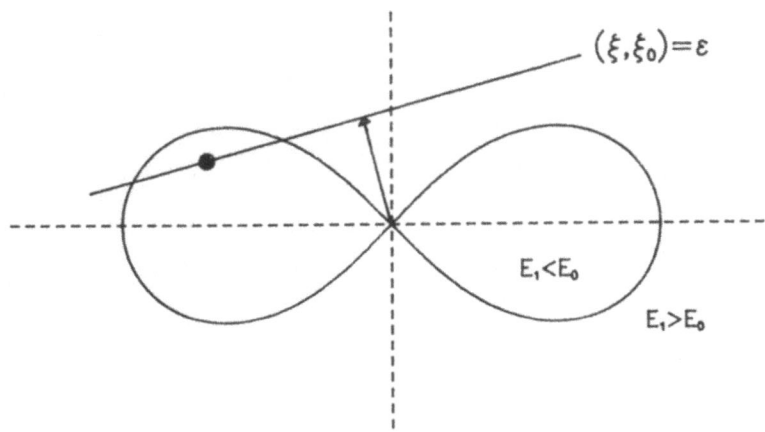

INSTABILITY FOUND BY SECOND MINIMIZATION OF E_1 WITH

RESPECT TO PERTURBATIONS ξ IN A HYPERPLANE

Figure 2. Energy landscape for nonlinear stability.

Our finite difference scheme is supposed to be second order accurate except perhaps at the magnetic axis $s = 0$ and the plasma surface $s = 1$. Therefore we extrapolate the growth rate ω to zero mesh size by assuming a representation of the form

$$- \omega^2 = A_0 + A_2 h^2 + A_3 h^3.$$

The coefficients A_0, A_2 and A_3 are determined numerically by a least squares fit to data for $- \omega^2$ computed for three or more choices of the mesh size h. Extensive calculations have established that this procedure works very well empirically (cf. Fig. 3). The value of $- \omega^2$ usually turns out to be positive for any mesh that occurs in practice, but as $h \to 0$ it decreases and may become negative in the limit, indicating instability. The apparent stability at

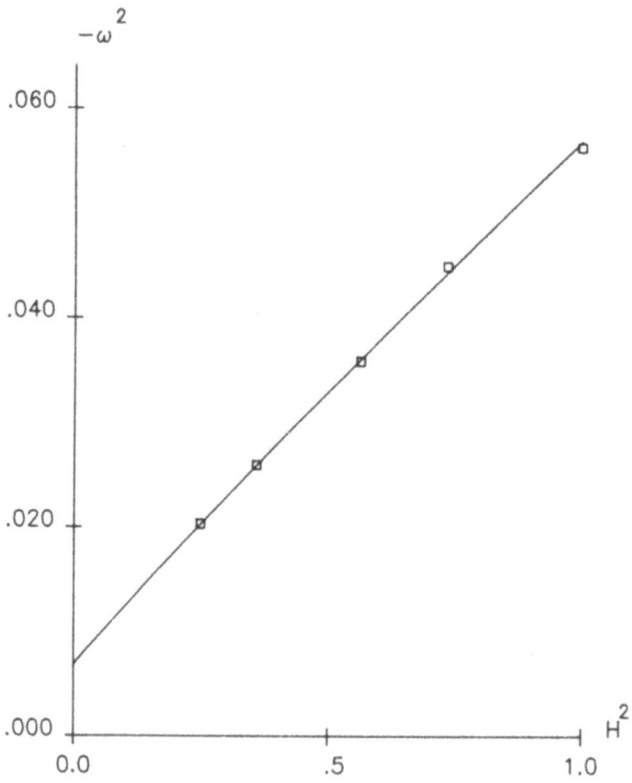

LEAST SQUARES EXTRAPOLATION TO ZERO MESH SIZE H
OF M=1 MODE OF HELIAC WITH A =8 , Q =3, C =1.6,
P=.08(1.−R²)², USING THE FORMULA −ω²=A₀+A₂H²+A₃H³
WITH H=1 CORRESPONDING TO 6X18X24 MESH CELLS

Figure 3. Convergence study of an $m = 1$ growth rate.

a fixed mesh may be attributed to the artificial viscosity inherent in our method. Also, there is virtually no destabilizing numerical resistivity because of our formulation of the flux constraints in terms of prescribed functions of s. We observe that equilibrium quantities such as the rotational transform μ or the net toroidal current I near the magnetic axis may be only first order accurate, so they must be estimated from a less optimistic representation in terms of h.

The test function ξ_0 is usually specified by the formulas

$$\delta R = 2s^{(m-1)/2}(1-s)\cos 2\pi(mu - nv),$$
$$\delta\psi = -s^{(m-2)/2}(1-2s)\sin 2\pi(mu - nv),$$
$$\delta r_0 = \delta z_0 = 0, \qquad \theta = 2\pi v$$

drawn from the example of a screw pinch, where $m \geq 2$ and n are poloidal and toroidal mode numbers [5]. When $m = 1$ these expressions are replaced by

$$\delta R = 2s^{1/2}(1-s)\cos 2\pi(u - nv),$$
$$\delta\psi = -(1-3s)\sin 2\pi(u - nv),$$
$$\delta r_0 = \cos 2\pi nv, \qquad \delta z_0 = \sin 2\pi nv.$$

Experience suggests that the simple trigonometric dependence of ξ_0 on the periodic coordinates u and v is adequate because the second minimization used to calculate E_1, which for $n = 1$ is performed over the full torus, automatically generates appropriate sidebands. However, at a resonant surface the dependence of a true unstable mode on s tends to be peakier. To approximate more complicated modes of that kind we may restrict the perturbation ξ to the interior of an inner flux tube that just encloses the resonance. Because of the rapid decay of growth rates in their dependence on the small radius of the plasma, this technique seems to simulate to some extent the effect of a radially peaked disturbance.

3. Comparison with Exact Solutions

In the case of a straight circular cylinder our stability analysis reduces to a one-dimensional energy principle described by Bateman [2]. Good agreement has been obtained between results calculated using our code and data based on that principle [5]. Similarly, growth rates produced by the three-dimensional code compare favorably with those computed by a two-dimensional spectral method developed for straight helically symmetric equilibria [13]. Such a comparison is plotted in Fig. 4.

In our calculation of growth rates there is a loss of significant figures associated with the subtraction of E_0 from E_1 and a further loss of significance in extrapolating to zero mesh size. For problems of physical interest as many as six figures may be lost altogether. Thus to have

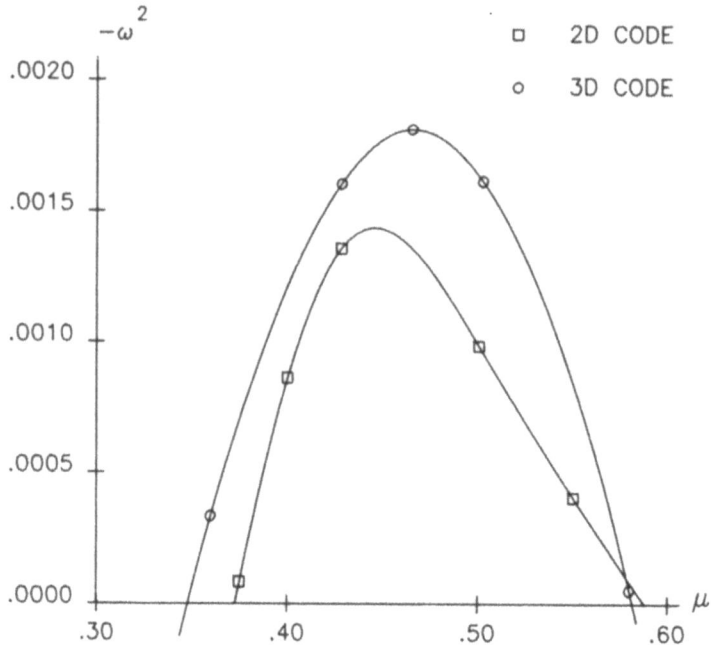

COMPARISON BETWEEN THE 3D CODE AND THE
HELICALLY SYMMETRIC VERSION OF THE ERATO
CODE FOR M=2, N=1 MODE WITH $.3 \leq \Delta_2 \leq .4$
AND β =.03, A =7.55, Q =5, P =$P_0(1.-R^2)$

Figure 4. Validation of global stability calculations.

confidence in the first significant figure, and therefore in the sign, of $-\omega^2$ one must often compute both E_0 and E_1 to an accuracy of seven significant digits on several grids. That it is possible to achieve this is primarily due to the absence of numerical resistivity in our method, which opens the way for an iterative scheme that converges to any desired degree of accuracy.

One must examine the dependence of $\delta W = E_1 - E_0$ on the amplitude ε_0 of the perturbation ξ to distinguish linear from nonlinear stability. It turns out that both can be studied for the lowest mode numbers m. Instabilities tend to saturate at amplitudes that do not destroy confinement. However, the corresponding bifurcated equilibria may well have magnetic islands that lead to anomalous transport [5]. Thus while the positivity of E_P suggests that some sort of equilibrium may exist in quite general circumstances, only solutions of the magnetostatic equations that are identified with nested surfaces and linear stability can be expected to result in good confinement.

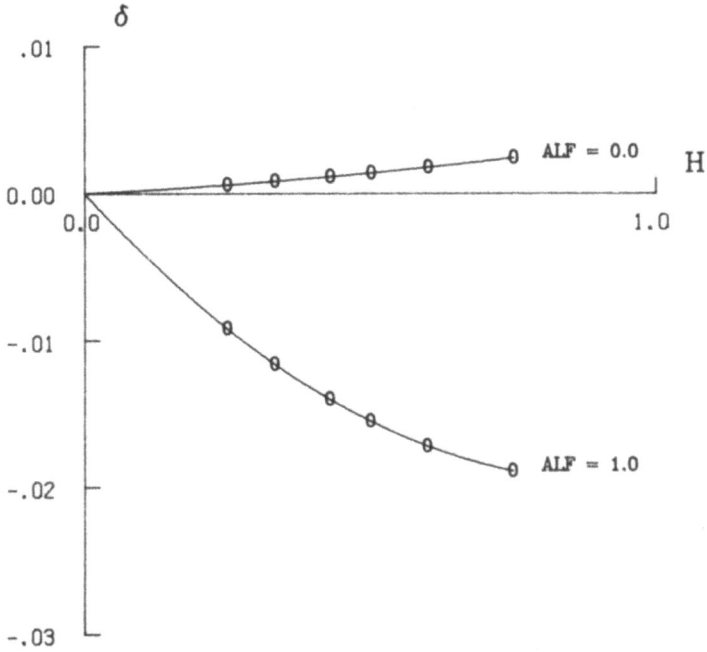

AXIS SHIFT δ VS H FOR AXIALLY SYMMETRIC
EQUILIBRIUM WITH H = 1 FOR 6×12 CELLS
AND WITH TWO RADIAL MESH SCALINGS ALF .

Figure 5. Convergence of the magnetic axis to an exact solution.

The bifurcated equilibria can be determined numerically by minimizing δW as a function of ε_0.

It is predominantly global stability that can be assessed by the procedure we have outlined. The theory of the continuous spectrum for one-dimensional solutions as well as calculations of linear stability for axially symmetric equilibria suggest that localized modes corresponding to large m and n are also important [34]. We shall discuss their stability in the next chapter, which is concerned with the Mercier criterion. Free surface modes can be dealt with, too, but the results are less decisive. Finally, one might attempt to superimpose a linearized stability analysis on the equilibrium calculation, but that is more tedious.

Convergence studies have also been made that compare the calculated shift δ of the magnetic axis for axially symmetric equilibria with known solutions of the tokamak problem (cf. Fig. 5). The results are only accurate to first order in the mesh size h.

CHAPTER 5

The Mercier Criterion

1. Local Mode Analysis

We have presented a method to test for instabilities with low mode numbers m and n, such as kink modes. Both theoretical analysis and calculations of axially symmetric equilibria suggest that modes with high m and n, including interchange modes, may be equally critical in the search for stable high pressure configurations [34]. They can be assessed via the Mercier criterion, which is concerned with modes localized about some rational surface [39]. The assumption is made that the plasma is covered by a family of nested flux surfaces, which we have seen to be rather tenuous when the geometry is truly three-dimensional [28]. Specialized variations δW of the energy E_P are introduced that are restricted to the neighborhood of a rational surface or a closed magnetic line. Then a necessary condition for stability is derived by careful optimization of δW. A preliminary version of this condition involves integrals over the relevant closed line, but considerations about the ergodicity of magnetic lines on a surface with irrational rotational transform lead to a better version involving surface integrals that occur naturally in our model of magnetohydrodynamic equilibrium. For details we refer to the book by Mercier and Luc [39], noting only that their derivation is somewhat at odds with the resonances that occur at rational surfaces in three dimensions. More specifically, the analysis suggests that in the neighborhood of a rational surface, lower energy levels can be achieved by equilibrium solutions that have island structure.

Let $V = V(s)$ be the volume enclosed by the toroidal flux surface $s = $ const., and use a prime $'$ to indicate differentiation with respect to the toroidal flux s, so that in particular

$$V' = \iint D \, du \, dv.$$

We formulate the Mercier criterion for local stability as an inequality

$$\Omega = \Omega_s + \Omega_w + \Omega_d \geq 0,$$

where

$$\frac{\pi^2 \mu^2}{s}\Omega_s = \frac{(\mu')^2}{4} - \mu' \iint \frac{(J - I'B) \cdot B}{(\nabla s)^2} D \, du \, dv$$

is a shear term including the stabilizing contribution $(\mu')^2$, where

$$\frac{\pi^2 \mu^2}{s}\Omega_w = p'\left[V'' - p' \iint \frac{D \, du \, dv}{B^2} \right] \iint \frac{B^2}{(\nabla s)^2} D \, du \, dv$$

is a term involving the magnetic well V'', and where

$$\frac{\pi^2 \mu^2}{s}\Omega_d = \left[\iint \frac{J \cdot B}{(\nabla s)^2} D \, du \, dv \right]^2 - \iint \frac{B^2}{(\nabla s)^2} D \, du \, dv \iint \frac{(J \cdot B)^2}{B^2 (\nabla s)^2} D \, du \, dv$$

is a combination of terms related to Schwarz's inequality for the Pfirsch–Schlüter current that is like the island width d introduced in Chapter 3. For a straight circular cylinder the Mercier criterion is known to reduce to Suydam's criterion

$$\frac{\pi^2}{s}\Omega = \frac{(\mu')^2}{4\mu^2} + \frac{p'}{R^2 B_3^3} \geq 0,$$

where B_3 stands for the toroidal magnetic field and the prime $'$ continues to indicate differentiation with respect to flux.

In the applications it is instructive to assess each of the terms Ω_s, Ω_w and Ω_d separately when examining the sign of Ω. In Ω_s the square $(\mu')^2$ is always stabilizing, but the influence of the second linear term depends on the sign of the shear μ'. It is desirable to have $V'' < 0$ so that the effect of the magnetic well on Ω_w is stabilizing for a standard pressure profile with $p' < 0$. Schwarz's inequality shows that the last term Ω_d is always unfavorable. In our model this can be due either to destabilizing Pfirsch–Schlüter currents or to the presence of magnetic islands that have destroyed the nested surfaces required in the construction of the equilibrium.

2. Computational Method

In the numerical evaluation of the Mercier criterion, which is related to that of the anomalous diffusion coefficient d^2, the most sensitive calculation is that of the scalar product

$$J \cdot B = \frac{1}{D}\left[B_2\left(\frac{\partial B_1}{\partial v} - \frac{\partial B_3}{\partial s} \right) + B_3\left(\frac{\partial B_2}{\partial s} - \frac{\partial B_1}{\partial u} \right) \right],$$

where use is made of the equilibrium condition

$$\frac{\partial B_2}{\partial v} - \frac{\partial B_3}{\partial u} = 0.$$

We refer to the listing of subroutine CDEN in Chapter 14 for details about the kind of differencing that is involved in a direct computation of $\nabla \times B$. Suffice it to say that our approach there is motivated by the quadrature formula used to approximate E_P.

An alternate procedure to calculate the Pfirsch–Schlüter current involves a version of the formulas in Section 3.2 that is suggested by the work of Boozer on transport [17, 18]. Because $J \cdot \nabla s = 0$, we can introduce Clebsch potentials ϕ and ζ such that

$$B = \nabla \phi + \zeta \nabla s$$

and

$$J = \nabla \zeta \times \nabla s.$$

More specifically, ζ is defined on each flux surface $s = $ const. by the path-independent integral

$$\zeta = \int \frac{p' d\phi - J \cdot B \, d\psi}{B^2}.$$

It follows that

$$\lambda = \frac{J \cdot B}{p' B^2}$$

can be found by integrating the first order partial differential equation

$$\frac{\partial \lambda}{\partial \phi} = -\frac{\partial}{\partial \psi} \frac{1}{B^2}.$$

This leads to a problem of small divisors near rational surfaces. However, because the coefficients of the differential equation are constant it can be solved in a practical way through Boozer's device of considering ϕ and ψ as the independent variables.

Both ϕ and ψ are readily available on each flux surface in our model because of the conservation form that is used in our difference scheme. We renormalize them to obtain variables

$$\tilde{\phi} = 2\pi\phi/I_P, \qquad \tilde{\psi} = 2\pi\psi - \mu\tilde{\phi}$$

that have the periods 2π, 0 and 0, -2π in the toroidal and poloidal directions, respectively, where I_P stands for the poloidal current. Here the assumption $I(s) \equiv 0$ is made, although the code allows for nontrivial net current. Let us consider the Fourier series expansion

$$\frac{1}{B^2} = \sum a_{mn} e^{im\tilde{\psi} + in\tilde{\phi}},$$

where it turns out that

$$a_{mn} = \frac{1}{(2\pi)^2} \iint \frac{e^{-im\tilde{\psi} - in\tilde{\phi}}}{B^2} d\tilde{\psi}\, d\tilde{\phi}$$

$$= \frac{1}{I_P} \iint e^{-im\tilde{\psi} - in\tilde{\phi}} D\, du\, dv,$$

since

$$B^2 = (\nabla s \times \nabla \psi)\cdot\nabla\phi = \frac{\partial(s, \psi, \phi)}{\partial(x, y, z)}.$$

These Fourier coefficients are relatively easy to compute in our code. Substitution into the partial differential equation for λ and formal term by term integration yields the expression

$$\lambda = -I_P \sum \frac{m a_{mn}}{n - \mu m} e^{im\tilde{\psi} + in\tilde{\phi}},$$

where the constant term drops out because $I'(s) = 0$. This expansion exhibits an expected resonance at rational $\mu = n/m$ that is connected with the nonstandard character of the equilibrium problem in magnetohydrodynamics [18, 28]. In practice we circumvent the difficulty by truncating the series for λ at values of the indices m and n falling short of resonance. Caution should be exercised when using apparently divergent series in this fashion, and cases with significant resonance at low m and n require special attention. In particular, a comparison with the direct calculation of $J \cdot B/B^2$ from $\nabla \times B$ is called for.

We have implemented both a direct scheme based on $\nabla \times B$ and an indirect scheme based on the Fourier series for λ in our evaluation of $J \cdot B/B^2$ for the Mercier criterion. The indirect scheme is preferred because it has a wider range of applicability. Experience has shown that the numerical results for Ω can be extrapolated to zero mesh size following rules similar to the one we introduced for $-\omega^2$ in Chapter 4 (cf. Fig. 6). In fact, the Mercier quantity Ω and the function $-\omega^2$ obey similar scaling laws.

In practice it is found that while data for Ω are inaccurate near the magnetic axis $s = 0$ and the plasma boundary $s = 1$, there is a middle range including the flux surface $s = 1/2$ where the extrapolated results are reliable. For many examples of physical interest it turns out that they are less optimistic about issues of stability than our previous estimates of global growth rates. However, they are considerably less expensive to compute. Comparisons with exact solutions for screw pinches and tokamaks have

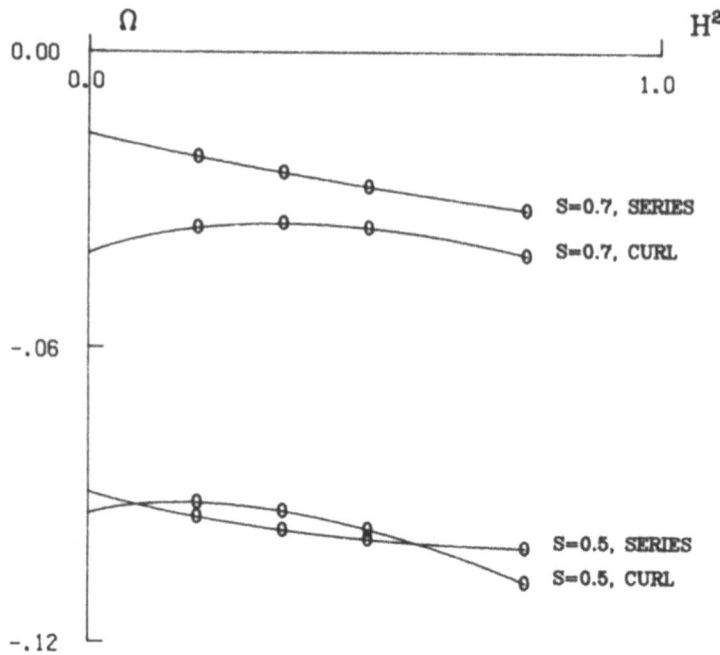

MERCIER CRITERION VS H FOR ATF—1

WITH β = .013, Δ_2= .25, Δ_3= −.05,

H=1 FOR 7×14×14 CELLS .

Figure 6. Convergence study for the Mercier criterion.

been made successfully. Finally, let us suggest that if a specific Fourier coefficient

$$b_{mn} = -\frac{I_p m a_{mn}}{n - \mu m}$$

in the expansion of λ is responsible for negative values of Ω, then modes with corresponding m and n ought to become dangerous. On the other hand, there is an innate contradiction between the derivation of the Mercier criterion and the divergence of the Fourier series for λ at rational surfaces in three dimensions.

Numerical results show that for most stellarator configurations the dominant term in the Mercier criterion arises from the $m=1$, $n=0$ term in λ. This is the Pfirsch–Schlüter current due to the toroidal effect, which appears with an amplification factor of μ^{-1}. Since a typical value of μ over one field period is usually small, that factor becomes significant. Low order resonances with $n \neq 0$ can also be important. However, if an assumption of helical symmetry is made, these terms drop out.

3. Island Width and Anomalous Diffusion

Our evaluation of the Mercier stability criterion $\Omega \geq 0$ for nested surface equilibria leads to speculation about the physical significance of the unfavorable inequality $\Omega < 0$ when it occurs. We have found harmonic stellarator fields with current density $J = 0$ and pressure $p = 0$ such that our calculation gives $J \cdot B \neq 0$ because of the presence of small magnetic islands and ergodic regions intermingled with the toroidal flux surfaces. Schwarz's inequality shows that the contribution Ω_d to Ω becomes negative in such cases when computed from $\nabla \times B$. Dimensional analysis then suggests that we interpret $-\Omega_d$ as an anomalous diffusion coefficient measuring the adverse effect on transport of the islands and ergodic regions. A similar transport coefficient computed using λ appears in the work of Grad on classical diffusion [29]. Furthermore, experience with our code and with other magnetohydrodynamic calculations indicates that linearly unstable local modes are usually nonlinearly stable and tend to saturate, producing bifurcated equilibria with resonant islands. The resulting losses may fall short of disruption even though $\Omega < 0$. These considerations lead us to propose the Mercier quantity

$$\eta = \max[0, -\Omega]$$

as an anomalous diffusion coefficient measuring transport associated with linear instabilities, resonance and poor magnetic surfaces. The possibility of confining plasma reasonably well for positive values of η below some critical average level is consistent with experimental data for β in the ISX-B tokamak and the Wendelstein W VII-A stellarator, where theoretical β limits have been exceeded. In a similar way the Mercier requirement $\Omega \geq 0$ is relevant to the diffusion in a core of plasma around the magnetic axis for tokamaks with μ near unity there.

Setting aside speculation about the physical significance of positive values of η, we turn our attention to numerical errors in the calculation of Ω. These are much smaller when the computation is performed using the Fourier series for λ in preference to $\nabla \times B$. After extrapolation to zero mesh size the values of Ω become consistent with other estimates of the Mercier criterion in simple cases. On a comparative basis it is reasonable to presume that a significant increase in Ω with β indicates stability, whereas a systematic decrease is destabilizing. However, anomalies in the derivation and evaluation of the Mercier criterion in three dimensions suggest caution in interpreting negative values of Ω alone as final evidence of instability.

CHAPTER 6
Transport

1. Classical and Neoclassical Models

Problems of diffusion and transport are just as important in plasma physics as questions of equilibrium and stability. Thus it is of interest to investigate how our construction of magnetohydrodynamic equilibria in three dimensions can be coupled with an analysis of diffusion. One approach is the method of classical transport developed by Grad and Hogan through omission of inertia terms in the partial differential equations of magnetohydrodynamics [30]. This suggests using averaged equations to track in physical time the poloidal flux, mass and entropy as functions of the toroidal flux s identified with a moving family of nested surfaces [29]. Our model of equilibrium can be used in such a calculation if the entropy is introduced appropriately and if care is taken to achieve good accuracy [8]. The diffusion coefficients that occur in the theory are not unlike the surface integrals appearing in the Mercier criterion. It might therefore be helpful to use the procedure described in Chapter 5 to compute the parallel component of current. The perpendicular component is readily found from the formula

$$(J \times B) \times B = p' \nabla s \times B.$$

We observe that a difficulty might be encountered in applying our code to Grad's model if the island width is appreciable so that there is significant anomalous diffusion $d^2 > 0$ at $p = 0$.

Since it is neoclassical rather than classical transport theory that appears to be most relevant to plasma confinement, a more immediate problem is to determine particle orbits and drift surfaces for the equilibria of our model. To that end Marcal has introduced a convenient divergence-free representation

$$B = \nabla \chi_1 \times \nabla u + \nabla \chi_2 \times \nabla v$$

of both the plasma and the vacuum fields that are calculated by the equilibrium and stability code [38]. This enables him to trace magnetic lines and reassess the nested surface hypothesis. He has applied the same technique to the guiding center equations

$$\frac{1}{\rho_\parallel} \dot{X} = B + \nabla \times \left(\rho_\parallel B \right)$$

of Boozer, in which the parallel gyroradius $\rho_\parallel$ is a specified function of the field strength [17]. This theory can be combined with Monte Carlo calculations to arrive at predictions about the particle confinement time of stellarator configurations with complicated geometry. Here a fundamental assumption is made that transport can be estimated for finite β by tracking the orbits of isolated particles in the magnetic field produced by our equilibrium code.

2. Flux Coordinates

The treatment of neoclassical transport that fits most effectively with our equilibrium model is that developed by Boozer in terms of the flux coordinates s, ψ and ϕ. His theory shows that guiding center orbits and particle transport can be assessed by Monte Carlo methods requiring only a knowledge of the Fourier series for $1/B^2$ and λ as functions of ϕ and ψ that we calculated in Section 5.2 (cf. [17, 20]). The Fourier coefficients a_{mn} that occur are output by the code in a form convenient for such an application. Resonance at a rational surface with low m and n is relatively easy to identify in this computation. A corresponding enhancement of transport is to be expected.

It becomes apparent that issues about transport have a connection with the question whether boundary value problems are well posed that purport to determine quantities like p, ζ or λ as periodic solutions of differential equations along the magnetic lines of force on a toroidal flux surface. This is the problem of small divisors encountered in the KAM theory [16, 41]. The calculation of λ from Boozer's Fourier decomposition of $1/B^2$ that we have exploited seems to be especially effective here because it introduces a partial differential equation with constant coefficients whose spectral solution emerges in closed form. Also, the question of nonexistence of equilibrium becomes more transparent in flux coordinates.

An analogy can be drawn between classical and neoclassical transport through a consideration of the partial differential equation

$$\left[B + \nabla \times \left(\rho_\parallel B \right) \right] \cdot \nabla \sigma = 0$$

for guiding center drift surfaces $\sigma = $ const. Putting $\sigma = s + \lambda_\parallel$ we find that

$$\frac{\partial \lambda_\parallel}{\partial \phi} + \cdots = -\frac{\partial \rho_\parallel}{\partial \psi},$$

where the dots stand for negligible terms involving products of the perturbations $\lambda_\parallel$ and $\rho_\parallel$. Thus $\lambda_\parallel$ has a relationship to $\rho_\parallel$ that is similar to the relationship of the Pfirsch–Schlüter current λ to the reciprocal field strength $1/B^2$. Large Fourier coefficients of λ may be symptomatic of variations in $\lambda_\parallel$ big enough to cause appreciable transport. Resonance of multiple harmonics can be expected to have a pronounced effect on $\lambda_\parallel$ when the gradient of $\rho_\parallel$ becomes large, as in the case of trapped particles, but changes in the sign of $\rho_\parallel$ then complicate matters further. The subtle influence of ergodic orbits may be as significant physically as the more apparent dependence of transport on mirror effects and on random changes in $\rho_\parallel$ due to collisions. In any numerical analysis of these phenomena it is important to retain the divergence-free character of the drift velocity field (cf. [38]).

For circulating particles we can interpret $\lambda_\parallel / |\nabla s|$ as a guiding center step size that has the same order of magnitude as the parallel gyroradius $\rho_\parallel$ on any flux surface that is not resonant. The scaling of difference approximations to the heat equation

$$f_t = f_{xx}$$

with the ratio of mesh sizes $(\Delta x)^2 / \Delta t$ and the construction of solutions by a random walk indicate that the diffusion coefficient for transport of such particles ought to be proportional locally to $\nu_e \lambda_\parallel^2 / (\nabla s)^2$, where ν_e is an effective collision frequency (cf. [40]). The theory of random evolution [42] suggests that a global transport coefficient can be introduced on each flux surface $s = $ const. by taking a mean value with respect to kinetic energy and magnetic moment of the square integral of $\lambda_\parallel / |\nabla s|$ with respect to the poloidal and toroidal variables ψ and ϕ. Denoting this average by $\|\lambda_\parallel\|^2$, we arrive at a semiempirical expression of the form

$$\tau = \int \frac{ds}{\nu_e \|\lambda_\parallel\|^2}$$

for the particle confinement time τ.

In the case of trapped particles, changes in the sign of $\rho_\parallel$ along any banana orbit lead us to use for ν_e the square of the bounce frequency divided by the collision frequency rather than the collision frequency itself. Observe that $\lambda_\parallel$ has the same relationship to the more familiar longitudinal adiabatic invariant

$$\zeta_\parallel = \int \rho_\parallel \, d\phi$$

that the Pfirsch–Schlüter current λ has to the Clebsch potential ζ specifying the current density J, namely,

$$\lambda_{\parallel} = - \frac{\partial \zeta_{\parallel}}{\partial \psi}.$$

For trapped particles what we propose is to multiply this by the ratio of the bounce frequency to the collision frequency, which ought to be done before averaging to form a transport coefficient. It may also become necessary to minimize the norm of $\lambda_{\parallel}$ with respect to an additive constant of integration that depends on ψ alone.

The influence of collisions is included by performing the mean value with respect to kinetic energy and magnetic moment over an appropriate proba- bility distribution F. Note that the norm $\|\lambda_{\parallel}\|^2$ essentially reduces to Grad's classical transport coefficient [29] in the limiting case of high collision frequency and small mirror ratio, where we can replace $\rho_{\parallel}$ by $1/B^2$. When resonance is negligible, our equilibrium and stability code provides an approximate calculation of the guiding center step size $\lambda_{\parallel}/|\nabla s|$ from the Fourier series with respect to $\tilde{\psi}$ and $\tilde{\phi}$ of

$$\rho_{\parallel} = \pm \rho_0 \sqrt{1 - \mu_0 B} / B,$$

where ρ_0 and μ_0 are parameters specifying the gyroradius and the magnetic moment. The relationship

$$\lambda_{mn} = - \frac{I_P m \rho_{mn}}{n - \mu m} = - \frac{m \rho_0}{n - \mu m} \int \int B \sqrt{1 - \mu_0 B} \, e^{-im\tilde{\psi} - in\tilde{\phi}} D \, du \, dv$$

between the Fourier coefficients λ_{mn} and ρ_{mn} of the periodic functions $\lambda_{\parallel}$ and $\rho_{\parallel}$ may explain how resonance can produce the anomalous transport that is observed experimentally [40]. In this context it is significant that ρ_{mn} can be expected to increase as one approaches the regime of trapped particles. Incidentally, with μ_0 set equal to its expected value $2/3$ the confinement time τ provides a measure of the failure of existence or local stability for toroidal equilibria in three dimensions.

The broader conclusion of all the transport theories seems to be that confinement time varies monotonically with the mirror ratio

$$T = \frac{\max|B| - \min|B|}{\max|B| + \min|B|}$$

on individual magnetic surfaces and with a transport coefficient of the form

$$- \frac{\Omega_{\lambda}}{(p')^2} = \frac{s}{\pi^2 \mu^2} \left\{ \int \int \frac{B^2 D}{(\nabla s)^2} \, du \, dv \int \int \frac{\lambda^2 B^2 D}{(\nabla s)^2} \, du \, dv - \left[\int \int \frac{\lambda B^2 D}{(\nabla s)^2} \, du \, dv \right]^2 \right\}$$

suggested by Grad's work and the Mercier criterion. With λ replaced by $\lambda_{\parallel}$, a similar quantity occurs in our definition of the geometric confinement time

$$\tau = \int \left[\int \frac{\nu_e}{I_P} \int \int \frac{\lambda_{\parallel}^2 B^2 D \, du \, dv}{(\nabla s)^2} \, dF(\mu_0, \rho_0) \right]^{-1} ds.$$

The difference $\Omega_d - \Omega_\lambda$ between the island width term Ω_d obtained from $\nabla \times B$ and the Pfirsch–Schlüter term Ω_λ obtained by formal integration of the Fourier series for $1/B^2$ becomes a suggestive measure of island width for $p' > 0$ that may be indicative of anomalous transport. We output the Fourier coefficients of $1/B^2$ and λ as well as the three quantities T, Ω_d and Ω_λ so that the code can be used to assess tradeoffs between the stability and transport properties of various equilibria. It has been found that solutions of the magnetostatic equations exhibiting multiple harmonics can have desirable properties in this respect. The Wendelstein W VII-AS stellarator device designed at the Max Planck Institute for Plasma Physics is a case in point [1]. More detailed results may be obtained by a Monte Carlo calculation.

As our discussion of confinement time suggests, the questions of equilibrium, stability and transport are all interrelated. More specifically, higher modes that are linearly unstable usually saturate at finite amplitude to create bifurcated equilibria with magnetic islands that correspond to the resonance of the mode. These islands in turn may enhance the anomalous transport associated with nonexistence of nested magnetic surfaces. Therefore, as we have pointed out in Section 5.3, the size of a quantity such as $\eta = \max[0, -\Omega)$ or τ may well serve as a measure of the confinement that should be expected in a given configuration.

It is interesting to observe that for a stellarator with net current $I = 0$, toroidal field coils introduce primarily omnigenous terms in the Fourier expansion of $1/B^2$, that is, terms which depend on ϕ alone. Our theory suggests that their effect on particle transport is not great. On the other hand, for a tokamak the corresponding terms that model toroidal field coils depend on $\phi + I\psi$ and are not omnigenous. In fact, the ripple associated with them enhances transport by an amount apparently proportional to the square of the net current I.

CHAPTER 7
Stellarator Experiments

1. The Mathematical Model

Toroidal geometry is advantageous for the confinement of a plasma by a magnetic field. If net current $I \neq 0$ is used not only to heat the plasma but also to maintain a poloidal field compensating for outward toroidal drift, then the geometry can be axially symmetric and the partial differential equations for equilibrium are two-dimensional. This is the case of the tokamak, which can be treated by mathematical methods simpler than ours unless one wants to study more abstruse questions such as bifurcated equilibria or bundle divertors. There are advantages, however, in maintaining zero net current $I = 0$, as in a stellarator or torsatron. Then steady operation becomes feasible in a reactor context. In a stellarator the poloidal field balancing toroidal drift is induced by helical coils, so that the magnetic surfaces have complicated three-dimensional geometry. Ideal magnetohydrodynamics, and more specifically our model based on the variational principle, offers one of the few theoretical tools available to analyze these more elaborate configurations adequately.

Helical coil winding laws can be studied by means of the sharp boundary version of our code that includes a vacuum field surrounding the plasma. However, we prefer to discuss first the case of a plasma that is confined inside a flux surface of prescribed shape. A convenient form for this boundary surface is defined in the modified cylindrical coordinate system of Chapter 2 by the formula

$$r_1 + iz_1 = r_b e^{2\pi i u} + \Delta_1 e^{2\pi i v} - \Delta_2 e^{-2\pi i(u-v)},$$

where

$$r_b = 1 - \Delta_0 \cos 2\pi v - \Delta_3 \cos 2\pi (3u - v) + \Delta_{10} \cos 2\pi u + \Delta_{20} \cos 4\pi u$$
$$+ \Delta_{30} \cos 6\pi u + \Delta_{22} \cos 4\pi (u - v) + \Delta_{33} \cos 6\pi (u - v).$$

The coefficients Δ_1, Δ_2 and Δ_3 are associated with $l = 1$, $l = 2$ and $l = 3$ helical coils, respectively. The remaining factors Δ_{jk} model the effect of current in coils with other winding laws corresponding to higher harmonics. The major radius of the plasma column is A and the average minor radius is 1, so that A specifies the aspect ratio. Any number of field periods Q is allowed, and straight configurations are included by putting the parameter $\varepsilon = 1/A$ equal to zero.

We shall be primarily concerned with stellarator equilibria having zero net current

$$I = I(s) \equiv 0,$$

which are calculated by iterating to determine the corresponding rotational transform $\mu = \mu(s)$. However, μ is held fixed during the second minimization of E_p used to analyze stability. Usually we start from a pressure distribution of the form

$$p = p_0(1 - s)^2$$

and put $\gamma = 2$, which serves to initialize the distribution of mass. In the study of stellarator configurations it is of interest to examine the dependence on the plasma parameter β of the rotational transform μ, the average shift

$$\delta = \int r_0 \, dv$$

of the magnetic axis, the island width d, the sign of the Mercier parameter Ω, and the eigenvalues $-\omega^2$ for resonant modes with n/m in the range of μ.

The analysis of the Pfirsch–Schlüter current

$$B \cdot \nabla \frac{J \cdot B}{B^2} = (\nabla p \times B) \cdot \nabla \frac{1}{B^2}$$

in Section 3.2, together with the requirement $I = 0$, implies that with the above choice of the pressure distribution p we have

$$J = 0$$

at the plasma surface $s = 1$. Indeed, because $p' = 0$ there ergodicity of the magnetic lines of force combines with $I = 0$ to show that $J \cdot B = 0$, and the perpendicular component of J vanishes, too. This means that at $s = 1$ the Mercier stability criterion reduces to

$$\Omega = \frac{(\mu')^2}{4\pi^2\mu^2} \geq 0$$

and is obviously satisfied. Moreover, by asking that B remain continuous across the plasma boundary we are led to a Cauchy problem for the elliptic system of partial differential equations

$$\nabla \times B = 0, \qquad \nabla \cdot B = 0$$

to determine a vacuum field surrounding the plasma. Optimistically this suggests that coils may be found to confine the plasma in its equilibrium without introducing either surface current or current density at $s = 1$. In such a situation the variational principle of magnetohydrodynamics guarantees that neither the free surface modes nor any localized modes at $s = 1$ can become unstable. Thus our choice of the pressure distribution is such as to transfer all questions of instability to an analysis in the interior of the plasma where the accuracy of our computational method turns out to be better.

2. The Heliotron E

The most ambitious stellarator experiment in the world today is the Heliotron E at Kyoto University [46]. In our theory we model it by a configuration with plasma aspect ratio $A = 10, Q = 18$ field periods (rather than the laboratory value $Q = 19$, which is not divisible by 2), and a pair of pure $l = 2$ coils yielding $\Delta_2 = 0.3$ at the plasma surface. For vanishing net current Heliotron E has the remarkable range

$$0.5 \leq \mu \leq 2.0$$

for the rotational transform, which straddles the dangerous rational surface $\mu = 1$. Because of the large rotational transform we find that the toroidal shift δ of the magnetic axis remains below 0.5 for β as high as 0.05. Thus there appears to be no difficulty with the existence of equilibrium in this range.

For Heliotron E a dangerous instability is the Kruskal–Shafranov kink mode with $m = 1, n = 1$. In preliminary calculations based on the test functions ξ_0 specified in Chapter 4 this mode appeared to become critical at $\beta = 0.05$ or thereabouts [7]. However, later computations based on perturbations ξ that are restricted to an inner flux tube barely enclosing the rational surface $\mu = 1$, and which consequently better simulate a peaky mode, suggest that the internal $m = 1, n = 1$ mode has a critical value of β closer to 0.02 (cf. Fig. 7). Note that $-\omega^2 > 0$ corresponds to a stable mode. This result agrees well with predictions made by Strauss and Wakatani [45, 46]. On the other hand, our calculation of the Mercier criterion for Heliotron E gives a significantly lower critical β (cf. Fig. 8). Thus far in the actual experiment $\beta = 0.018$ has been achieved after reduction of impurities in the laboratory plasma.

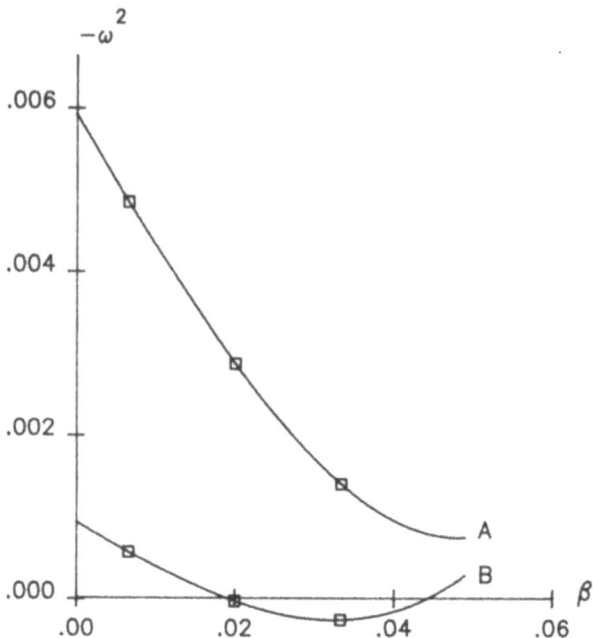

M=N=1 MODE OF HELIOTRON E WITH Q=18

A : PLASMA SURFACE A=10, Δ_2=.30,

   $P=P_0(1.-1.0R^2)^2$

B : INNER FLUX TUBE A=13, Δ_2=.27,

   $P=P_0(1.-0.6R^2)^2$

Figure 7. Radial refinement of a Heliotron mode.

3. The Wendelstein W VII-A

The other major stellarator experiment now in operation is the Wendelstein W VII-A at the Max Planck Institute for Plasma Physics in Garching. It has an $l = 2$ winding law with $Q = 5$ field periods and an aspect ratio $A = 20$ for the plasma. The high aspect ratio leads to such small growth rates that our numerical estimates of the difference of energies $\delta W = E_1 - E_0$ lose significance and become unreliable. However, the values of $-\omega^2$ for the global $m = 2$, $n = 1$ mode associated with the resonant surface $\mu = 1/2$ do suggest a critical value of β not exceeding 0.01. This is consistent with measurements

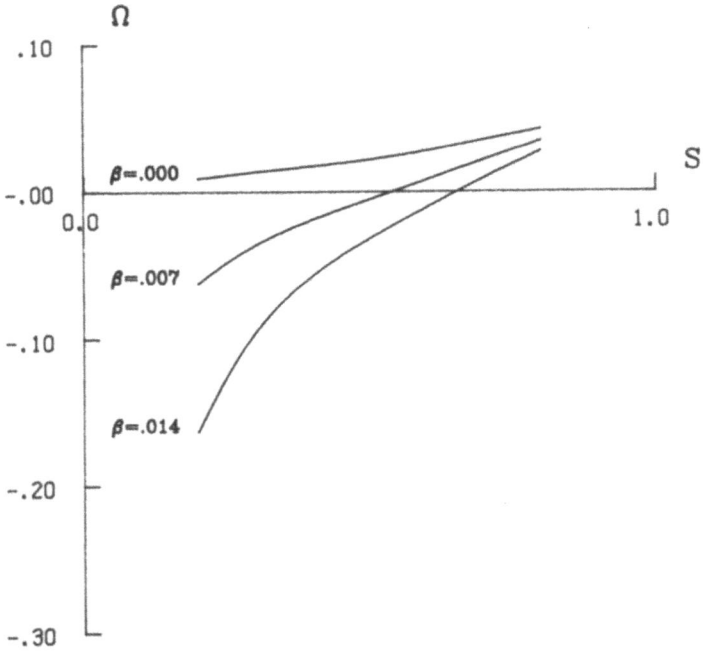

MERCIER CRITERION FOR HELIOTRON E WITH

$A=10.0$, $Q=18$, $\Delta_2=0.3$, $P=P_0(1.-R^2)^2$.

Figure 8. Local instability of the Heliotron E.

from the present experiment, which has achieved an average β of 0.005. Expansions in the neighborhood of the magnetic axis [39] predict that an $l=2$ stellarator is unstable to local modes for a pressure profile of the form $p = p_0(1-s)^2$. Our calculations of the Mercier criterion for Wendelstein W VII-A with $\beta = 0.007$ yield similar results (cf. Fig. 9). In contrast, a flatter profile of the form $p = p_0(1-s^2)^2$ becomes stable near the axis and is only slightly unstable over the rest of the plasma region.

As β increases, the magnetic axis shifts outwards and the plasma digs its own magnetic well. At the same time, the Pfirsch–Schlüter current increases and contributes a destabilizing term. Moreover, small sidebands created by the interaction of the helical and toroidal fields may be enough to change the sign of Ω near a resonant surface. In general, it is simple to study stability trends with parameter changes, but it is difficult to decide how negative Ω must be before an instability will be observed experimentally.

Our earliest proposal for a stellarator that might reach a critical β for both equilibrium and stability as high as 0.03 was an $l=2,3$ configuration

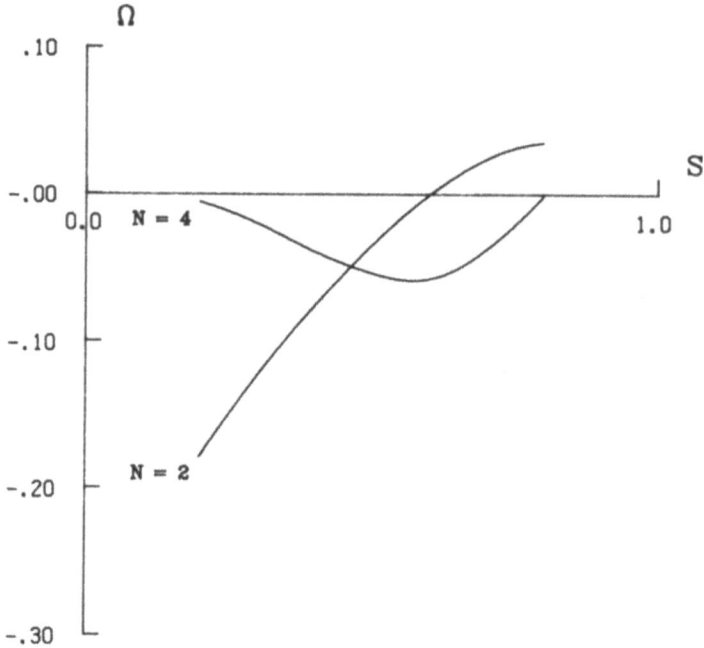

MERCIER CRITERION FOR W VII–A

WITH $\beta = .007$, $\Delta_2 = .35$,   A = 20,

Q=5, $P = P_0(1.- R^N)^2$.

Figure 9. Local instability of the Wendelstein W VII-A.

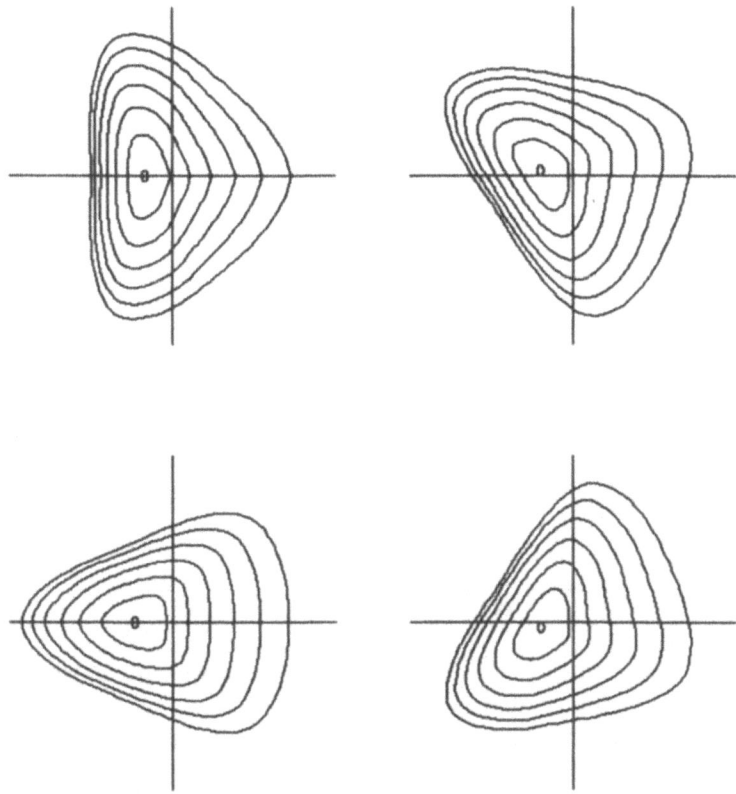

CROSS SECTIONS AT V= .00,.25,.50,.75, 1/(EP•QLZ)= 0.83

MAJOR RADIUS= 10.00   MINOR RADIUS= 1.00

Figure 10. Shift of the magnetic axis in an $l = 2, 3$ field.

with $A = 10, Q = 12$, and with precisely two harmonics $\Delta_2 = -\Delta_3 = 0.15$ different from zero [4]. In analogy with high β stellarator theory, the multiple harmonics serve to shift the magnetic axis significantly inward at $\beta = 0$, so that when β increases the outward shift will not become fatal (cf. Fig. 10). Stability properties of the gross $m = 1$, $n = 1$ and $m = 2$, $n = 1$ modes of the $l = 2, 3$ stellarator appeared to be satisfactory despite the presence of a magnetic hill at $\beta = 0$ associated with the inward shift of the magnetic axis. However, as β increases the rotational transform develops more and more shear and eventually breaks out of the realistic range $0 \leq \mu \leq 1$. In fact, at $\beta = 0.03$ the value of μ at the magnetic axis becomes negative, indicating the possible appearance of an $m = 2, n = 0$ island that might be damaging to confinement. Because of the magnetic hill $V'' > 0$, Mercier's criterion predicts poor stability properties of the $l = 2, 3$ stellarator.

CHAPTER 8

The ATF-1 and the WISTOR-U

1. Helical Coil Winding Laws

Neither of the two large American stellarator experiments, namely the C stellarator at Princeton that went out of operation in 1968 and the high β Scyllac at Los Alamos that was terminated in 1977, were very successful. This was in part because of unawareness of the mathematical subtlety of the equilibrium and stability problems that arise in the design of such devices. Several sounder proposals for stellarator or torsatron experiments have recently been under consideration in the United States. In drawing up the new proposals use has been made of three-dimensional equilibrium and stability codes [3, 22, 43], and extensive calculations have been performed to trace magnetic field lines and particle orbits as well as to assess particle transport by Monte Carlo methods [20, 36]. Here and in Chapter 9 we shall describe some predictions about the proposed configurations that follow from an application of our equilibrium and stability code.

To start with let us consider how to model a helical coil winding law by means of the sharp boundary version of the code. In the vacuum region surrounding the plasma we represent the magnetic field B as the gradient of a scalar potential ϕ satisfying Laplace's equation. The boundary values of ϕ are prescribed on an outer control surface Γ. In the exterior of Γ let us introduce another harmonic function ϕ^* which is single-valued and has the same normal derivative as ϕ on Γ, so that

$$\frac{\partial \phi^*}{\partial \nu} = \frac{\partial \phi}{\partial \nu},$$

where ν is the unit normal. The corresponding jump of $B \times \nu$ across Γ is a surface current

$$K = (\nabla\phi - \nabla\phi^*)\times\nu$$

that is parallel to the level curves of $\phi - \phi^*$. In the case of a stellarator ϕ has no period in the poloidal direction around the torus Γ because there is no net toroidal current I. Consequently these level curves become closed circuits that can be used to specify modular coils creating the stellarator field under discussion [1].

In a cruder approximation we can conceive of the level curves of ϕ itself on Γ as describing the shape and location of the helical coils. In this formulation of the problem we introduce the expression

$$\phi = c_1 v + c_2 \sin 2\pi u + c_3 \tan^{-1}\frac{c_4 \sin w}{1 - c_4 \cos w}$$

for the prescribed boundary values of ϕ, where

$$w = 2\pi v - l(2\pi u - \alpha \sin 2\pi u).$$

The coefficients c_1 and c_2 specify the principal toroidal field and an auxiliary vertical field, respectively. Similarly the parameters c_3 and c_4 specify the strength and excursion of helical coils. The more precise form of the helical coil winding law is characterized by the level curves of the function w, which depends on a modulation parameter α introduced originally in connection with the concept of an ultimate torsatron [26]. As usual Q indicates the number of field periods, and l is the number of helical coils. Sharp boundary equilibrium calculations establish that these parameters can be varied systematically to control the Fourier coefficients Δ_j and Δ_{jk} occurring in the formula for the plasma surface given in Section 7.1.

The University of Wisconsin proposed an experiment called the WISTOR-U which has an ultimate torsatron winding law with $Q = 12, l = 2$ and $\alpha = 0.63$ (cf. [7, 36]). For this configuration we have used sharp boundary, vacuum field calculations to show that the gross $m = 1, n = 1$ Kruskal–Shafranov free surface mode is stable for $\beta \leq 0.03$. These calculations, together with field line tracing data, indicate that the plasma surface of the WISTOR-U can be approximated by the formula of Section 7.1 with $A = 10$, $\Delta_1 = 0.2$, $\Delta_2 = 0.2$ and $\Delta_3 = -0.1$, but with all other Fourier coefficients set equal to zero. In this model of the WISTOR-U a second stability region develops for the global $m = 1$, $n = 1$ and $m = 2$, $n = 1$ modes before any instabilities are reached [7]. This is explained by the observation that an outward shift of the magnetic axis causes the plasma to dig a significant magnetic well as β increases (cf. [43]). However, predictions for the WISTOR-U using the Mercier local stability criterion are more pessimistic.

2. Beta Limits for the ATF-1

An ATF-1 experiment similar to the WISTOR-U is under construction at the Oak Ridge National Laboratory [43]. In that configuration we have again $Q = 12$ and $l = 2$, but $\alpha = 0$ and $A = 7$. In the formula for the plasma surface we put $\Delta_2 = 0.25$ and assume at first all other Fourier coefficients to vanish. At low β there is more shear than for the WISTOR-U, with μ in the range $0.3 \leq \mu \leq 0.9$. However, as β increases the distribution of rotational transform flattens out, with the value at the edge of the plasma falling off significantly (cf. Fig. 11). If the strength of the poloidal field is insufficient there may be a resonance of the $m = 2, n = 1$ mode. As in the case of the WISTOR-U, a second stability region for this mode is attained safely provided the strength of the $l = 2$ harmonic is kept at $\Delta_2 = 0.25$ (cf. Fig. 12). However, this prediction is inconsistent with calculations of the Mercier criterion for local modes, which show the stability β limit of ATF-1 with $\Delta_3 = 0$ to be under 0.013 (cf. Fig. 13).

Instabilities associated with the Mercier criterion can sometimes be stabilized by introducing a vertical field that shifts the magnetic axis outward. In general this augments the magnetic well, but of course it also reduces the equilibrium limit on β associated with the shift. In our model we simulate the vertical field by adding a term in Δ_3 to the equation of the plasma surface. This is motivated by the relationship between the shift and the product $\Delta_2\Delta_3$ that was mentioned in Section 7.3. Figures 14 and 15 display the dependence of the Mercier criterion and the eigenvalue $-\omega^2$ of the $m = 3, n = 2$ mode on Δ_3 for this model of the ATF-1. The choice $\Delta_3 = 0.05$ seems to stabilize both the global $m = 3, n = 2$ mode and the Mercier criterion. It should be noted that details of the pressure profile may change the results in the neighborhood of the axis. On the other hand, greater instabilities are found when $\Delta_3 = -0.05$, so that the magnetic axis is shifted inwards. The ATF-1 experiment has been designed with a capability for testing such predictions.

We observe an outward shift of the magnetic axis greater than half the plasma radius for $\beta = 0.04$, so we advance that as an equilibrium limit for the ATF-1 (cf. Fig. 16). This is more pessimistic than calculations carried out at the Oak Ridge National Laboratory using a flux-conserving model and a generalization of the Chodura–Schlüter code based on the spectral method [43]. However, those results refer to a more conventional choice of the pressure distribution. As for transport, we estimate that the mirror ratio is $T = 0.3$ at $s = 1$.

Without a vertical field to shift the magnetic axis outward and create a magnetic well, it is seen that the stability limit of the ATF-1 experiment falls below 0.01. With the vertical field included stability seems assured, but the outward shift of the magnetic axis becomes excessive for $\beta \geq 0.03$. More-over, when $\beta = 0.03$ the rotational transform flattens out so that all shear is

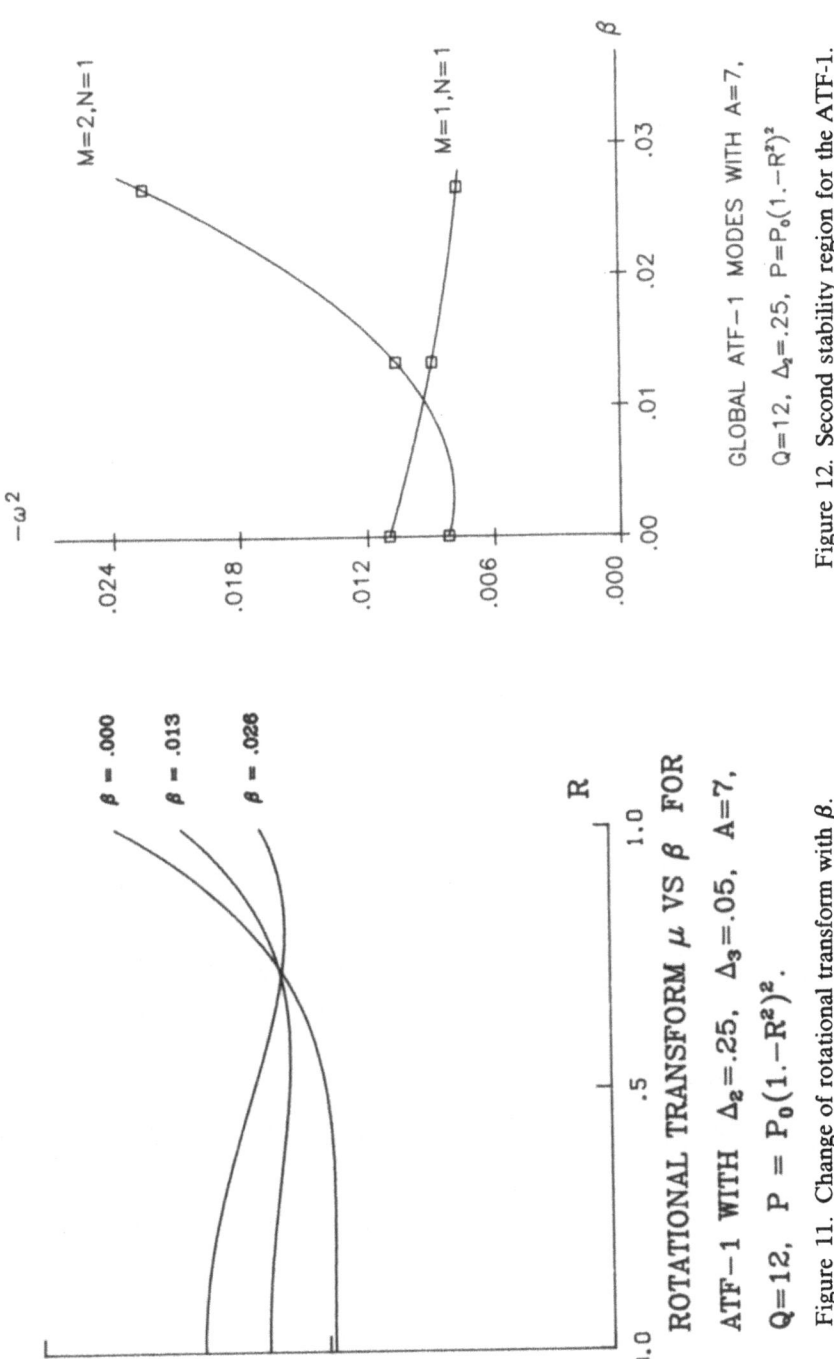

GLOBAL ATF−1 MODES WITH A=7,

$Q=12$, $\Delta_2=.25$, $P=P_0(1.-R^2)^2$

Figure 12. Second stability region for the ATF-1.

ROTATIONAL TRANSFORM μ VS β FOR

ATF−1 WITH $\Delta_2=.25$, $\Delta_3=.05$, A=7,

$Q=12$, $P = P_0(1.-R^2)^2$.

Figure 11. Change of rotational transform with β.

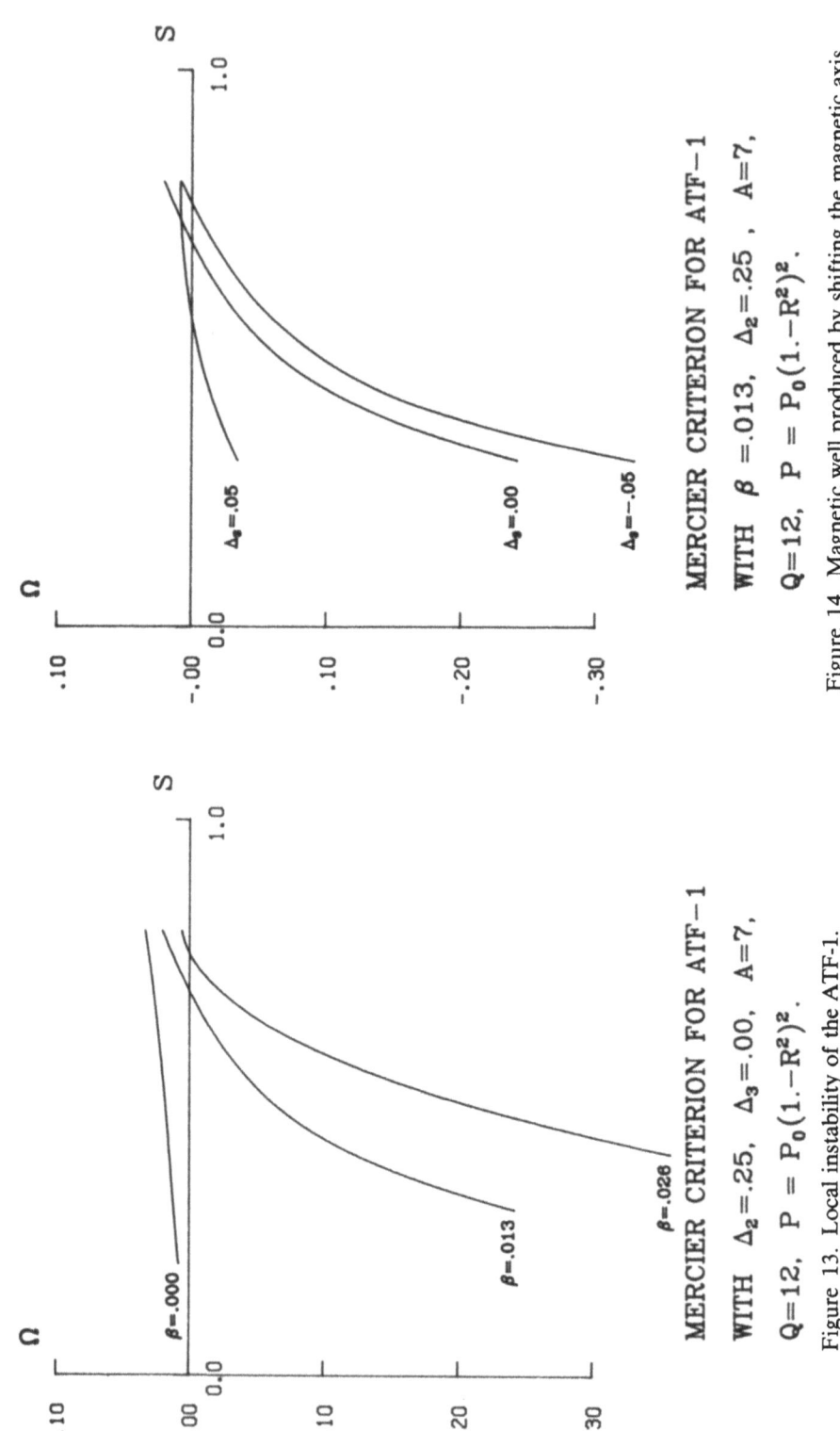

MERCIER CRITERION FOR ATF−1

WITH $\beta = .013$, $\Delta_2 = .25$, $A = 7$,

$Q = 12$, $P = P_0(1.-R^2)^2$.

Figure 14. Magnetic well produced by shifting the magnetic axis.

MERCIER CRITERION FOR ATF−1

WITH $\Delta_2 = .25$, $\Delta_3 = .00$, $A = 7$,

$Q = 12$, $P = P_0(1.-R^2)^2$.

Figure 13. Local instability of the ATF-1.

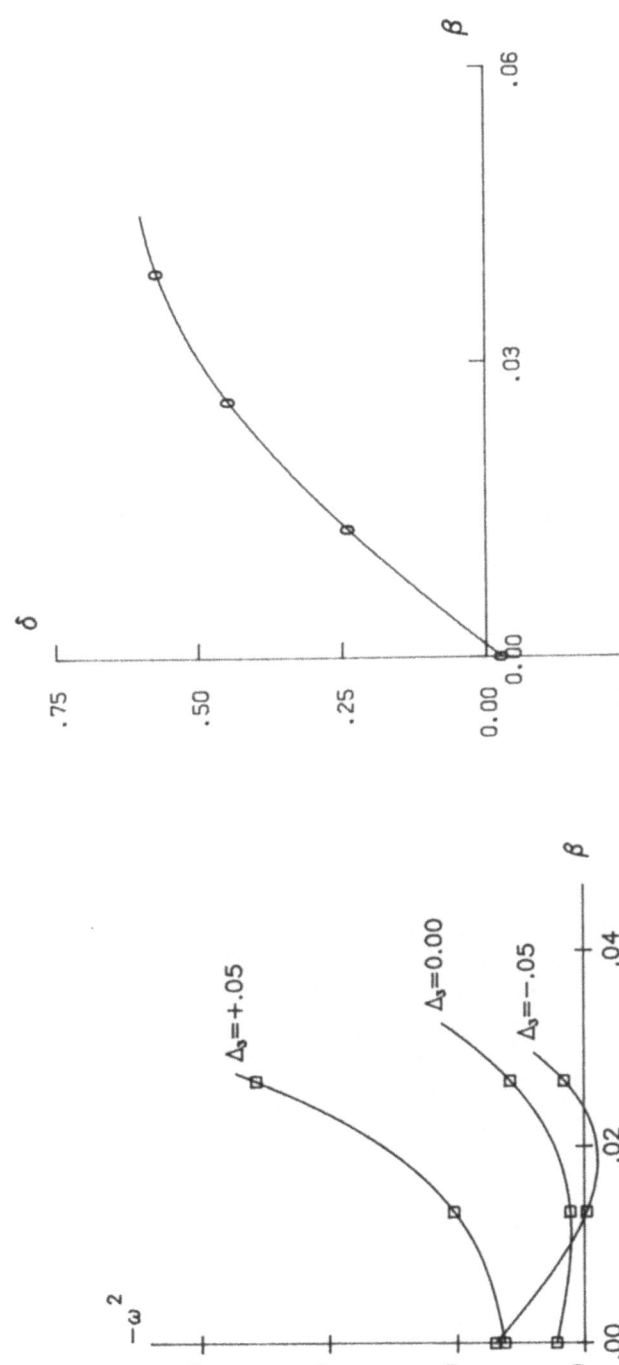

AXIS SHIFT δ VS β FOR ATF-1

WITH $\Delta_2 = .25$, $\Delta_3 = .00$, $A = 7$,

$Q = 12$, $P = P_0(1. - R^2)^2$.

Figure 16. Equilibrium β limit of the ATF-1.

DEPENDENCE ON Δ_3 OF M=3,N=2 MODE FOR ATF-1

WITH $\Delta_2 = .25$, $A = 7$, $Q = 12$, $P = P_0(1. - R^2)^2$

Figure 15. Stabilization of ATF-1 by a vertical field.

lost. This leads to resonance that may break up the magnetic surfaces and cause transport to deteriorate. We therefore predict an overall β limit for the ATF-1 of 0.03.

We have been unable to find stellarator configurations with critical β above 0.03 for both equilibrium and stability when the rotational transform is restricted to the interval

$$0 < \mu < 1.$$

However, the success of the Heliotron E experiment in achieving stable laboratory plasmas with μ straddling the resonant value $\mu = 1$ suggests consideration of rotational transforms in a higher range such as

$$1 < \mu < 2.$$

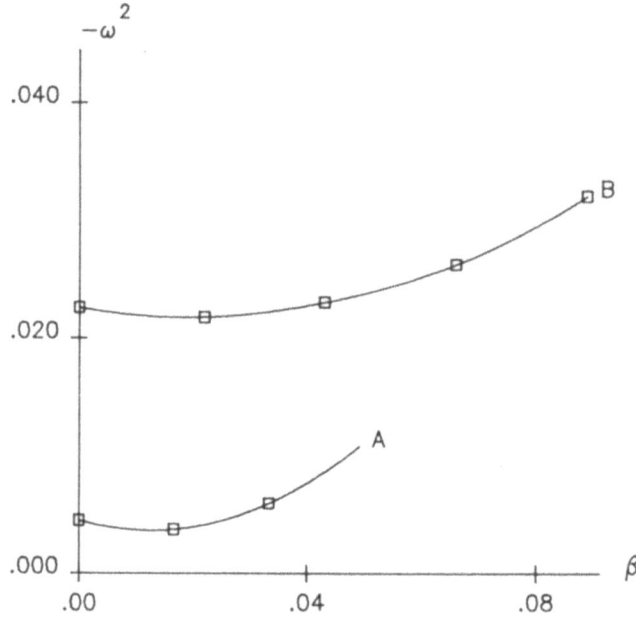

M=2 MODES WITH A=15 AND MU > 1

FOR A :   WISTOR−U−II AND B :   HELIAC

A :   Q=18, N=3, Δ_1=.22, Δ_2=.27, Δ_3=−.09

B :   Q= 6, N=4, C=1.6, Δ_A=.43, Δ_B=1.0, Δ_2^*=.67

Figure 17. Global stability of WISTOR-U-II and Heliac.

With this in mind we have investigated a new configuration, to be referred to as the WISTOR-U-II because of its connection with the Wisconsin proposal, that may have an equilibrium β limit near 0.08. It is defined by adding field periods to the WISTOR-U and putting $Q = 18$, $A = 15$, $\Delta_1 = 0.22$, $\Delta_2 = 0.27$, $\Delta_3 = -0.09$ in the plasma surface formula, which corresponds roughly to a modulation parameter $\alpha = 0.5$ in the ultimate torsatron winding law.

The results presented in Fig. 17 show that the global $m = 2$, $n = 3$ mode of the WISTOR-U-II stellarator goes directly into the by now familiar second stability region before developing any instability for finite β. That conclusion has been reinvestigated by examination of the Mercier criterion, which unfortunately predicts a low stability β limit for the WISTOR-U-II. However, the outward shift of the magnetic axis is significantly less than that for the original WISTOR-U because of the relatively high range $1.2 \leq \mu \leq 1.8$ obtained for the rotational transform.

CHAPTER 9

Heliac

1. Helically Symmetric Equilibria

Straight equilibria with helical symmetry have been investigated extensively as a source of promising stellarator configurations. In our earliest work on the variational method they have led us to unduly optimistic conclusions about the stabilizing effect of an $l = 3$ harmonic $\Delta_{33} \neq 0$ on high β stellarators [3]. That misconception was the outgrowth of surprising resolution of the first version of our code in the case of the Scyllac experiment at Los Alamos, which showed it to be unstable to the most dangerous $m = 1$, $n = 0$ mode even on the crudest meshes.

More recent studies at the Max Planck Institute for Plasma Physics have produced helically symmetric equilibria identified with judicious choices of the coefficients $\Delta_{11} = \Delta_1$, Δ_{22} and Δ_{33} that have critical β as high as 0.05 (cf. [35]). However, the optimal configuration of this kind seems to be the Heliac revived at the Princeton Plasma Physics Laboratory after having lain dormant for many years [19, 23]. Two-dimensional models of the Heliac have been shown to be stable for β up to 0.2. In this chapter we shall ascertain how toroidal effects alter the situation with regard to stability. We shall also apply the vacuum version of our three-dimensional code to develop evidence for the existence of the required Heliac equilibria in the toroidal case and to arrive at a practical modular winding law.

2. Stability

For the surface of the Heliac plasma we introduce the formula

$$r_1 + iz_1 = \left[\Delta_A + \Delta_B\left(1 + \Delta_2^* \cos 2\pi u\right)e^{iC\sin 2\pi u}\right]e^{-2\pi iv}$$

suggested by work at Princeton [19]. We have implemented this in subroutine SURF, taking into account that the poloidal coordinate u now
vanishes on a curve that twists once the short way around the torus in each
field period. To apply a severe test of stability we consider a set of
parameters $\Delta_A = 0.43$, $\Delta_B = 1$, $\Delta_2^* = 0.67$ such that for $Q = 3$ field periods of
the device the rotational transform is close to the most dangerous resonant
value $\mu = 1$. Both the $m = 1$, $n = 1$ and the $m = 2$, $n = 2$ global modes have
been calculated in such configurations using the test function ξ_0 described in
Section 4.2. In Fig. 18 we present results for two different values $C = 1.6$ and
$C = 1.0$ of the crescent parameter C which indicate that both the effect of
larger C and the effect of positive toroidicity $\varepsilon = 1/A$ are stabilizing. The
data in Fig. 19 show that even for the $m = 1$, $n = 1$ mode in the straight case
$\varepsilon = 0$ there is a second stability region that may perhaps be associated with
an outward helical shift of the magnetic axis. Figure 17 displays a comparison of the dependence of growth rate on β for stable, resonant $m = 2$ modes
of the WISTOR-U-II and Heliac with plasma aspect ratio $A = 15$. Both
examples have rotational transform in the interval $1 < \mu \le 2$.

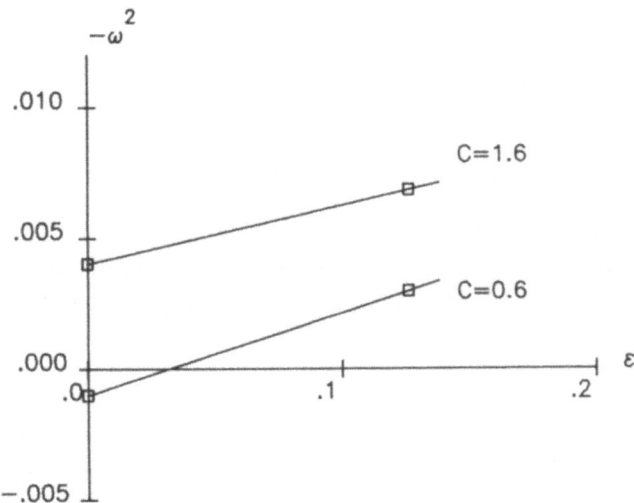

M = 1 MODE OF HELIAC WITH β = .04 IN

DEPENDENCE ON INVERSE ASPECT RATIO $\varepsilon = 1/A$

FOR Q= 3, Δ_A=.43, Δ_B=1.0, Δ_2^*=.67

Figure 18. Toroidal stabilization of Heliac.

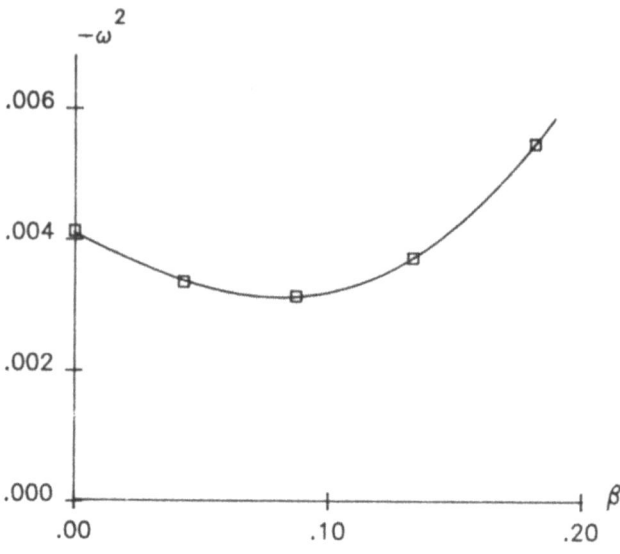

M=1, N=1 MODE OF HELIAC WITH

$\varepsilon=0$, Q=3, C=1.6, $\Delta_A=.43$, $\Delta_B=1.0$,

$\Delta_2^*=.67$, $P=P_0(1.-R^2)^2$

Figure 19. Second stability region of straight Heliac.

We presented in Fig. 3 a convergence study of growth rates showing that the calculations of gross stability for Heliac are reliable. Moreover, for $\varepsilon = 0$ they are in agreement with helically symmetric computations by other methods [19]. A typical stability run of 12,000 artificial time cycles determining E_0 over one field period and E_1 over three field periods of Heliac on a grid with $12 \times 36 \times 144$ cells takes 2.5 hours on the CRAY computer.

Results of the Mercier local stability criterion for Heliac are appreciably less costly to obtain. However, their implications are substantially different and should perhaps not be taken at face value. The resonance at $\mu = 1/3$ over one field period becomes quite perceptible in the Fourier expansion of the Pfirsch–Schlüter current λ, and the more familiar toroidal effect on λ associated with $\mu = 0$ is also significant. These contributions to Ω_λ or Ω_d combine to make Ω negative near $s = 0$ and $s = 1$ over what might otherwise be considered a stable range of β (cf. Fig. 20). There is some improvement if the configuration is altered to remove the resonance at $\mu = 1/3$, $s = 0.8$, where islands might be expected to form (cf. Fig. 21). The negative values near $s = 0$ can be removed by considering a flatter pressure profile near the magnetic axis. Moreover, the magnitude of $\eta = \max[0, -\Omega]$ remains small

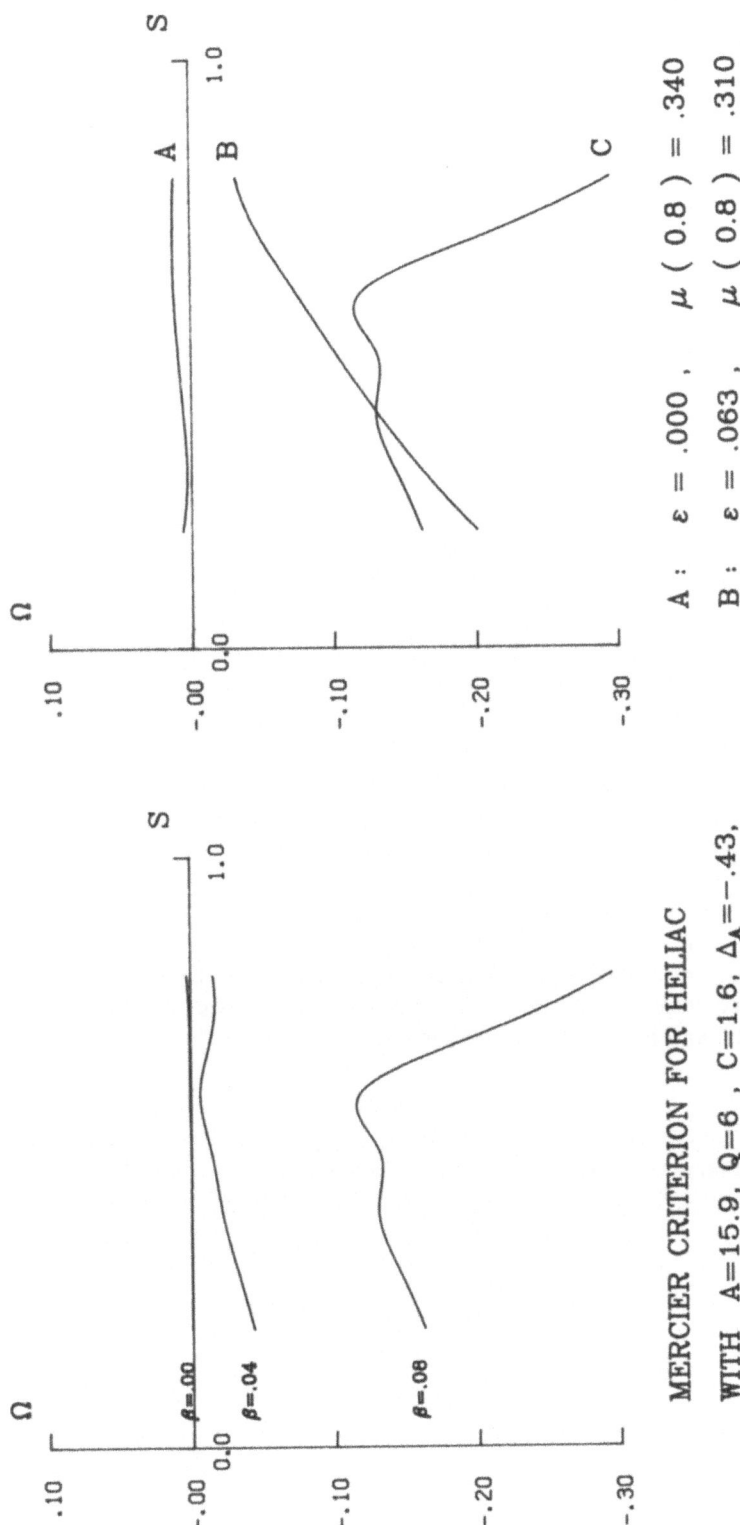

MERCIER CRITERION FOR HELIAC

WITH A=15.9, Q=6, C=1.6, $\Delta_A = -.43$,

$\Delta_B = 1.0$, $\Delta_2^* = 0.67$, $P = P_0(1. - R^2)^2$.

Figure 20. Resonance at $\beta = 0.08$.

A : $\varepsilon = .000$, $\mu (\ 0.8\ ) = .340$

B : $\varepsilon = .063$, $\mu (\ 0.8\ ) = .310$

C : $\varepsilon = .063$, $\mu (\ 0.8\ ) = .330$

Figure 21. Dependence of Ω on ε and μ at $\beta = 0.08$.

relative to the size of the values of β that are under investigation. We therefore conclude that even the pessimistic results of the Mercier criterion that we have obtained for Heliac suggest that it has a remarkable stability property compared to any other known stellarator configuration. In this context it should be borne in mind that multiple harmonics in genuinely three-dimensional equilibria may well make the Fourier series for λ or $\lambda_\parallel$ diverge if all the terms are kept. No comparable difficulty arises in cases with two-dimensional symmetry because the corresponding series depend on just one variable.

3. Free Boundary Equilibria

There is a controversy about the existence of Heliac equilibria in the toroidal case for high values of β. To investigate this question we have performed equilibrium calculations with the free boundary version of our code on grids with as many as 12 intervals of the radial coordinate s in the vacuum region and 12 intervals of s in the plasma region, combined with 36 intervals in the poloidal coordinate u and 48 intervals in the toroidal coordinate v. We have used the same formula for the control surface Γ that was introduced for the Heliac plasma surface, but with more moderate values $\Delta_A = -0.3$, $\Delta_B = 1.09$, $\Delta_2^* = 0.7$, $C = 1.2$ of the parameters so that the cross sections become convex (cf. Fig. 22). The boundary values of the scalar potential ϕ are defined by the expression

$$\phi = c_1 v + c_2 \sin 2\pi(u - v) + c_3 \tan^{-1}\frac{c_4 \sin 2\pi u}{1 + c_4 \cos 2\pi u} + c_5 \sin 2\pi u.$$

With $Q = 6$ field periods and a plasma aspect ratio $A = 15$ there appear to be satisfactory equilibria for $\beta \le 0.08$. This is consistent with calculations of the toroidal and helical shifts of the magnetic axis when the plasma surface remains fixed [19].

As the mesh is refined both the convergence and the resolution of the free boundary equation improve. However, it must be acknowledged that the free boundary version of the code is less satisfactory than the plasma version because we did not use derivatives of the Hamiltonian $E_P - E_V$ with respect to nodal values of the free surface function g in deriving finite difference equations for g. Our favorable results on the existence of high β Heliac equilibria contrast with negative evidence from runs of the Chodura–Schlüter code at the Oak Ridge National Laboratory, which diverge. We believe the divergence may be attributed to a failure to eliminate the magnetic lines as real characteristics of the pressure equation.

The formulas we have employed to calculate free surface Heliac equilibria suggest a simple modular winding law identified with the closed level curves of our boundary data for ϕ on the control surface Γ. Six or eight of these

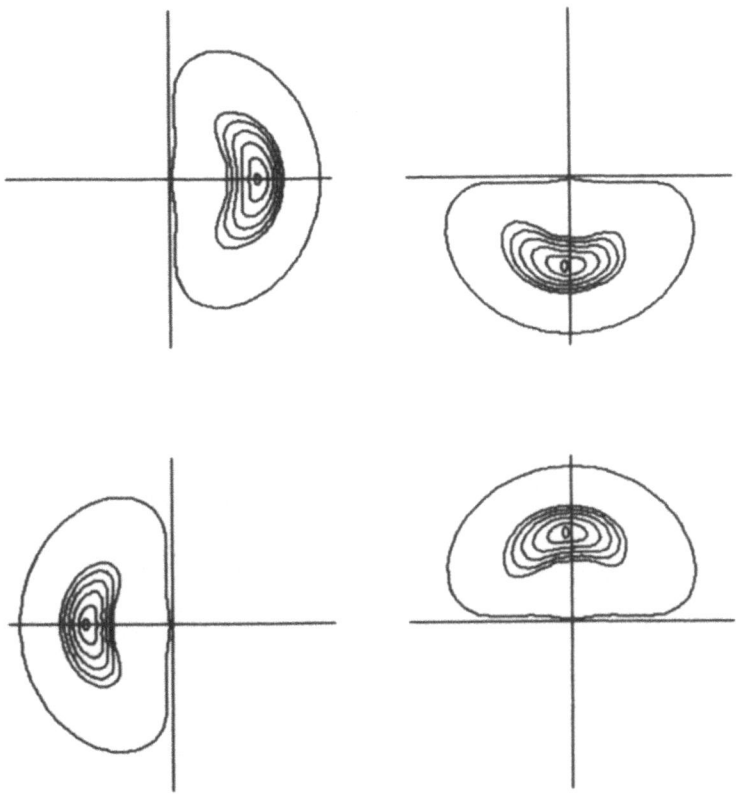

CROSS SECTIONS AT V= .00,.25,.50,.75, 1/(EP•QLZ)= 1.32

MAJOR RADIUS= 7.94   MINOR RADIUS= 1.00

Figure 22. Heliac equilibrium with a free boundary.

modular coils per field period might suffice to yield a satisfactory vacuum field. Runs of the three-dimensional code indicate that the shape and location of the plasma respond in an altogether reasonable fashion to variations in the parameters specifying the winding law, such as the vertical field strength c_2 or the inverse distance c_4 to singularities that can be associated with a poloidal field coil. Even more interesting, perhaps, is the dependence of the solution on the coefficient c_5, which is to be interpreted either as the current in spiral trimming coils that compensate for helical shift of the plasma, or as an equivalent tilt of the modular coils relative to the magnetic axis. The effect of c_5 on helical shift is displayed in Fig. 23, where $c_1 = 12.0$, $c_2 = -0.02$, $c_3 = 1.3$, $c_4 = 0.59$ and $c_5 = -0.2$.

The pronounced helical excursion of the Heliac plasma is reminiscent of bifurcated tokamak equilibria that we considered some time ago [3]. To illustrate the nonlinear capabilities of the code we exhibit a more refined

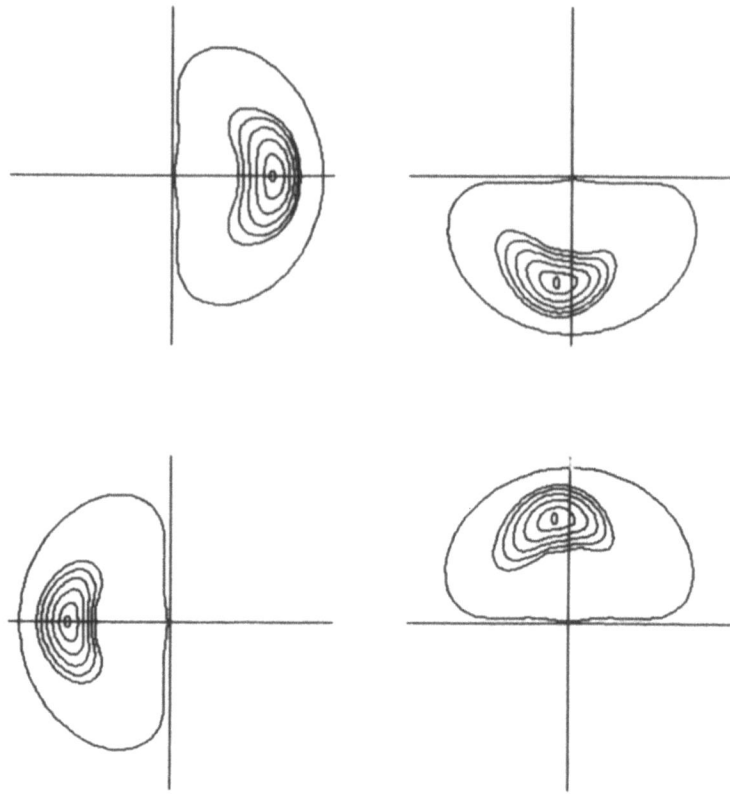

```
CROSS SECTIONS AT V= .00,.25,.50,.75, 1/(EP•QLZ)= 1.32
MAJOR RADIUS=  7.94                      MINOR RADIUS= 1.00
```

Figure 23. Effect of tilting modular coils.

calculation of this kind in Fig. 24. It brings to mind the original poloidal field coil winding law for Heliac that was proposed by the Princeton Plasma Physics Laboratory [19].

While the effect of sizeable distortions of the plasma such as occur in Heliac may be quite favorable for stability without tangible detriment to equilibrium, the implications for transport are less promising. In typical Heliac calculations of equilibrium in realistic toroidal geometry with $Q = 6$, $A = 15$, $C = 1.6$ and $\beta = 0.08$ we arrive at an estimate of $T = 0.15$ for the mirror ratio, but there is appreciable resonance. We hope to return to the discussion of transport for stellarators of this kind in more depth elsewhere.

According to the calculations that have been described here, Heliac seems to be the most attractive stellarator configuration to have been proposed up to this time. In our opinion it merits further theoretical and experimental investigation. More specifically, we are interested in a device with $Q = 6$

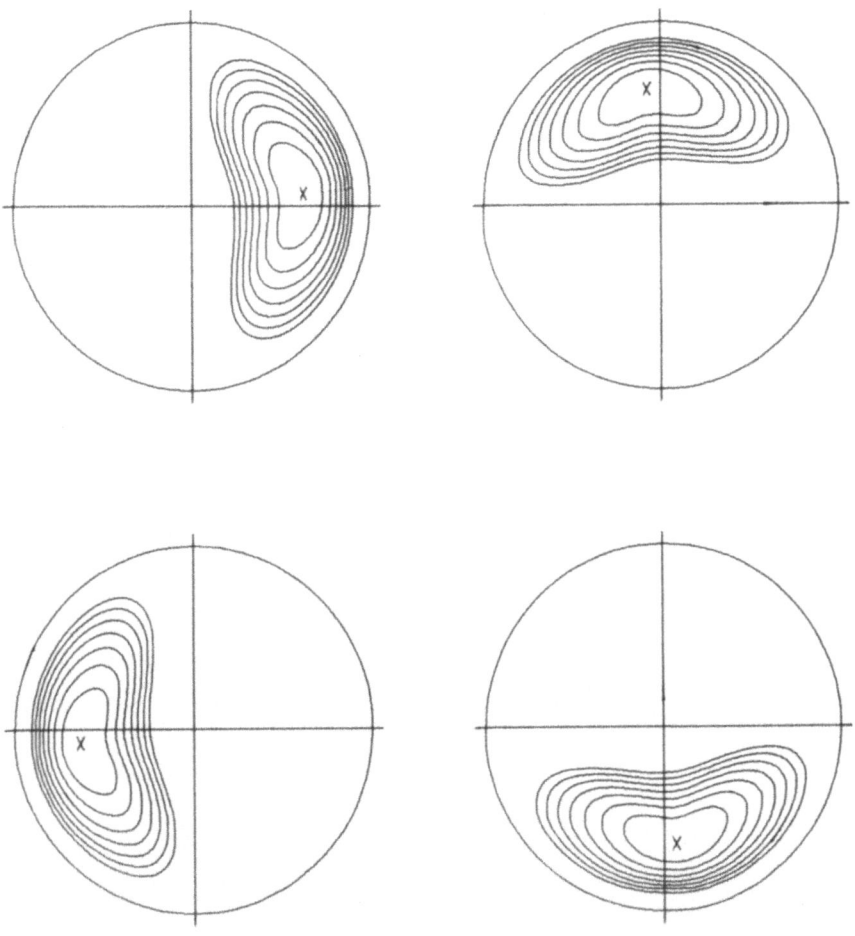

CROSS SECTIONS AT V= .02,.27,.52,.77, QLZ/2*PI= 1.50

MINOR RADIUS= 1.00

MAJOR RADIUS INFINITE

Figure 24. Bifurcated equilibrium of a screw pinch.

MODULAR COILS FOR HELIAC

WITH   A=3.3, Q=3 , C=1.2,

$\Delta_A=-.3,\ \Delta_B=\ 1.09,\ \Delta^*_2=\ 0.7$.

Figure 25. Three-dimensional geometry of a stellarator.

field periods, a plasma aspect ratio $A = 12$, and rotational transform near 1.6. This might be achieved using about 36 modular coils of the form

$$v = c' \tan^{-1} \frac{c'' \sin 2\pi u}{1 + c'' \cos 2\pi u}$$

wound on a control surface of aspect ratio $A = 6$ (cf. Fig. 25 for an example with $Q = A = 3$). The β limit ought to be at least 0.08 and transport may become acceptable. For the same gyroradius and collision frequency, this Heliac configuration has a value of the geometric confinement time τ introduced in Chapter 6 comparable to that of the Heliotron E and that of the Wendelstein W VII-A. Trimming coils should be included to shape and control the plasma, compensating for toroidal and helical shift. It is desirable that there be some shear, which is achieved by tightening the winding law. Finally, one might also consider a more conventional Heliac with rotational transform $\mu < 1$ and with less helical excursion of the magnetic axis, but with enough field periods to suppress low order resonance.

CHAPTER 10
References

1. *Application for Preferential Support for Wendelstein VII-AS*. Max Planck Institute for Plasma Physics, 1981.
2. Bateman, G. *MHD Instabilities*. Cambridge: MIT Press, 1978.
3. Bauer, F., O. Betancourt, and P. Garabedian. *A Computational Method in Plasma Physics*. Springer Series in Computational Physics. New York: Springer-Verlag, 1978.
4. Bauer, F., O. Betancourt, and P. Garabedian. "Numerical studies of new stellarator concepts." *J. Comp. Phys.*, 1980, *35*, 341–355.
5. Bauer, F., O. Betancourt, and P. Garabedian. "Nonlinear magnetohydrodynamic stability." *Phys. Fluids*, 1981, *24*, 48–54.
6. Bauer, F., O. Betancourt, P. Garabedian, and J. Shohet. "Beta limits for torsatrons." *Proc. Natl. Acad. Sci. U.S.A.*, 1981, *78*, 1–3.
7. Bauer, F., O. Betancourt, P. Garabedian, and J. Shohet. "Finite β stellarators." *I.E.E.E. Trans. Plasma Sci.*, 1981, *PS-9*, 239–243.
8. Bayliss, A., H. Grad, and O. Betancourt. "Numerical simulation of transport in tokamaks with ripple." Paper 2C19, Annual Controlled Fusion Theory Conference, Austin, 1981.
9. Berkowitz, J., H. Grad, and H. Rubin. "Problems in magnetohydrodynamic stability." *Proceedings of the Second United Nations International Conference on the Peaceful Uses of Atomic Energy (Geneva)*, 1958, *31*, 177–189.
10. Betancourt, O., and P. Garabedian. "Computer simulation of the toroidal equilibrium and stability of a plasma in three dimensions." *Proc. Natl. Acad. Sci. U.S.A.*, 1975, *72*, 926–927.
11. Betancourt, O., and P. Garabedian. "Equilibrium and stability code for a diffuse plasma." *Proc. Natl. Acad. Sci. U.S.A.*, 1976, *73*, 984–987.
12. Betancourt, O., and P. Garabedian. "Numerical analysis of equilibria with islands in magnetohydrodynamics." *Comm. Pure Appl. Math.*, 1982, *35*, 365–378.
13. Betancourt, O., F. Herrnegger, P. Merkel, J. Nührenberg, R. Gruber, and F. Troyon. "Comparison of MHD stability results obtained with the BETA 3D and HERA 2D codes." *J. Comp. Phys.*, 1983, *52*, 187–197.
14. Betancourt, O., and G. McFadden. "Nonparametric solutions to the varia-

tional principle of ideal magnetohydrodynamics." *Analysis, Geometry and Probability: Proc. of the First Chilean Symposium* Edited by R. Chuaqui. New York: Marcel Dekker, to be published.

15. Bloch, E. "Numerical solution of free boundary problems by the method of steepest descent." *Phys. Fluids (Suppl. II)*, 1969, *12*, 129–132.

16. Bogolyubov, N., J. Mitropoliskii, and A. Samoilenko. *Methods of Accelerated Convergence in Nonlinear Mechanics.* New York: Springer-Verlag, 1976.

17. Boozer, A. H. "Guiding center drift equations." *Phys. Fluids*, 1980, *23*, 904–908.

18. Boozer, A. H. "Plasma equilibrium with rational magnetic surfaces." *Phys. Fluids*, 1981, *24*, 1999–2003.

19. Boozer, A. H., T. K. Chu, R. L. Dewar, H. P. Furth, J. A. Gorec, J. L. Johnson, R. M. Kulsrud, D. A. Monticello, G. Kuo-Petravic, G. V. Sheffield, S. Yoshikawa, and O. Betancourt. "Two high-beta toroidal configurations: A stellarator and a tokamak-torsatron hybrid." *9th International Conference on Plasma Physics and Controlled Nuclear Fusion Research*, IEAE-CN-41/Q4, 1982.

20. Boozer, A. H., and G. Kuo-Petravic. "Monte Carlo evaluation of transport coefficients." *Phys. Fluids*, 1981, *24*, 851–859.

21. Brackbill, J. U. "Numerical magnetohydrodynamics for high-beta plasma." *Methods in Computational Physics*, 1976, *16*, 1–41.

22. Chodura, R., and A. Schlüter. "A 3D code for MHD equilibrium and stability." *J. Comp. Phys.*, 1981, *41*, 68–88.

23. Furth, H., J. Killeen, M. Rosenbluth, and B. Coppi, "Stabilization by shear and negative V''." *Plasma Physics and Controlled Nuclear Fusion Research*, 1965, *1*, 103–126.

24. Garabedian, P. R. "Analytic methods for the numerical computation of fluid flows." *Analytic Methods in Mathematical Physics*. Edited by R. P. Gilbert and R. G. Newton. New York: Gordon and Breach, pp. 169–184, 1970.

25. Gibson, A. "Trajectories of magnetic field lines in toroidal stellarators." *Phys. Fluids*, 1967, *10*, 1553–1560.

26. Gourdon, C., P. Hubert, and D. Marty. "Ultimate simplification of windings producing a magnetic configuration of the stellarator kind," *C. R. Acad. Sci. Paris*, 1970, *271*, 843–845.

27. Grad, H. "Some new variational properties of hydromagnetic equilibria." *Phys. Fluids*, 1964, *7*, 1283–1292.

28. Grad, H. "Toroidal containment of a plasma." *Phys. Fluids*, 1967, *10*, 137–154.

29. Grad, H. "Plasma transport in three dimensions." *Annals of the New York Acad. Sci.*, 1980, *357*, 223–235.

30. Grad, H., and J. Hogan. "Classical diffusion in a tokamak." *Phys. Rev. Lett.*, 1970, *24*, 1337–1340.

31. Greene, J. M. "Introduction to resistive instabilities." LRP 114/76, CRPP, École Polytechnique Fédérale de Lausanne, 1976.

32. Greene, J. M., and J. L. Johnson. "Determination of hydromagnetic equilibria." *Phys. Fluids*, 1961, *4*, 875–890.

33. Greene, J. M. and J. L. Johnson. "Hydromagnetic equilibrium and stability." *Advan. Theor. Phys.*, 1965, *1*, 195–244.

34. Grimm, R. C., J. M. Greene, and J. L. Johnson, "Computation of the magnetohydrodynamic spectrum in axisymmetric toroidal confinement systems," *Methods in Computational Physics*, 1976, *16*, 253–280.

35. Herrnegger, F. "Numerical studies of magnetohydrostatic finite-beta stellarator equilibria." *Z. Naturforsch.*, 1982, *37a*, 879–891.

36. Kruckewitt, T., and J. Shohet. "Magnetic properties of ultimate torsatrons." *Nucl. Fusion*, 1980, *20*, 1375–1380.

37. Kruskal, M. D., and R. M. Kulsrud. "Equilibrium of a magnetically confined plasma in a toroid." *Phys. Fluids*, 1958, *1*, 265–274.
38. Marcal, M. "Magnetic and drift surfaces in toroidal plasma equilibria." *Research and Development Report MF-98*, DOE/ER/03077-178, Courant Inst. Math. Sci., New York Univ., 1982.
39. Mercier, C., and H. Luc. *Lectures in Plasma Physics*. Commission of the European Communities, Luxembourg, 1974.
40. Miyamoto, K. *Plasma Physics for Nuclear Fusion*. Cambridge: MIT Press, 1976.
41. Moser, J. "On invariant curves of area-preserving mappings of an annulus." *Nachr. Akad. Wiss. Göttingen, Math. Phys. Kl.*, 1962, *IIa*, 1–20.
42. Papanicolaou, G. "Stochastic equations and their applications." *Amer. Math. Monthly*, 1973, *80*, 526–545.
43. *Proposal to Build ATF-1: An Advanced Toroidal Facility*, Oak Ridge National Laboratory, 1982.
44. Shohet, J. L. "Stellarators." *Fusion*, Vol. 1, Part A. Edited by E. Teller. Academic Press, pp. 243–289, 1981.
45. Strauss, H. R. and D. A. Monticello. "Limiting beta of stellarators with no net current." *Phys. Fluids*, 1981, *24*, 1148–1155.
46. Wakatani, M. "Interchange-ballooning mode in $l = 2$ helical systems." *I.E.E.E. Trans. Plasma Sci.*, 1981, *PS-9*, 243–247.

CHAPTER 11

Description of the Computer Code BETA

1. Introduction

The computer code BETA has been written to implement the solution of the magnetostatic equations discussed in the previous chapters. In this chapter an attempt is made to express the equations and parameters at least partially in the FORTRAN notation used in the code. This should help the user to understand and intercept the code whenever necessary. The equations were expressed in mathematical terms in the earlier chapters.

The procedure for running the code is relatively straightforward. All the necessary instructions appear in this chapter and in the glossary. However, an understanding of the theoretical framework is needed to determine the scope and limitations of the code and to interpret the results correctly. The code can only be used efficiently if the input parameters are chosen with care.

The program minimizes the potential energy, and if the corresponding iteration is convergent it computes an equilibrium. The iteration can diverge either because there is no equilibrium or because the equilibrium is unstable. However, if the iteration does diverge we must distinguish between numerical and physical instabilities. A poor choice of input parameters characterized by loss of the energy inequality DELENER < 0 may lead to numerical instabilities. Physical instabilities, on the other hand, are characterized by indefinite descent of the energy with the residuals of the the equilibrium equations increasing. When the iteration is convergent we must consider the effect of truncation error on the minimization process. The truncation error tends to make equilibria at a given mesh size appear more stable than they really are for the continuous equations. Therefore the mesh must be refined to see if the results remain unchanged. To study a given device, equilibrium and stability analyses are done for a set of increasingly finer meshes. In order to evaluate the stability, the resulting growth rates are extrapolated to zero mesh size using a polynomial least squares fit. All other variables should be extrapolated similarly to obtain meaningful results.

The computation consists of a two-dimensional axisymmetric calculation followed by a three-dimensional calculation that is divided into two main sections, (1) equilibrium together with the Mercier criterion and (2) global stability. The two-dimensional analysis is to be omitted by putting ASYE $\geq$ 100 if the equilibrium is three-dimensional. The code can be used for plasmas with a fixed boundary or with a free boundary surrounded by a vacuum region.

To clarify the notation we mention that it has been convenient in the code to replace R by $s^{1/2}R$ and ψ by $-\psi$. Also, we note that the derivative p' of the pressure has been evaluated in terms of V'' to improve accuracy.

2. Geometry

The code is initialized by an input file which defines the physical, geometrical and numerical parameters. That input data is stored on a filed called TAPE 25. The geometry of the case under investigation is the first input. The unit of length is equal to the average radius of the cross sections of the outer wall. In these units EP is the inverse of the major radius of the torus, which becomes the aspect ratio, and RBOU is the ratio of the free boundary radius to the wall radius. If no vacuum region is present RBOU = 1. We denote by QLZ the number of field periods into which the torus is divided. If EP > 0 the computation of the equilibrium solution is restricted to one such field period. For equilibrium the toroidal coordinate v varies over the interval $0 \leq v \leq 1$ in one field period. To perform an equilibrium computation over the full torus we must set QLZ = 1. If EP = 0 the major radius is infinite. This corresponds to the calculation of straight equilibria, and QLZ becomes the length of a field period.

The integer NRUN is the number of field periods for which a stability calculation can be performed. If NRUN = 1, equilibrium and stability are computed for just one period. If NRUN > 1 stability is computed for a domain consisting of NRUN periods, but the equilibrium input is given for one period. The code has been written so that after computing the equilibrium in one field period of the configuration, the calculation can be extended to several periods to carry out a stability analysis for modes with larger wave length.

The geometry of the outer wall is the next input. It is specified in subroutine SURF. The code has several representations for the wall which are selected by the value of NGEOM. For NGEOM = 1 we have the standard form given in Section 7.1, namely,

$$R(\overline{U},V) = RAD\ COS(\overline{U}) + DEL1\ COS(V) - DEL2\ COS(\overline{U} - V),$$

$$Z(\overline{U},V) = RAD\ SIN(\overline{U}) + DEL1\ SIN(V) + DEL2\ SIN(\overline{U} - V),$$

where

$$RAD = 1 - DEL0\ COS(V) + DEL10\ COS(\overline{U}) + DEL20\ COS(2\overline{U})$$

$$+ DEL30\ COS(3\overline{U}) - DEL3\ COS(3\overline{U} - V) + DEL22\ COS[2(\overline{U} - V)]$$

$$+ DEL33\ COS[3(\overline{U} - V)]$$

and

$$\overline{U} = U + ALFU\ SIN(U).$$

The zoning parameter ALFU is used to obtain a better distribution of mesh points so that the grid becomes evenly spaced in the poloidal angle.

If all the DEL's are zero, R and Z describe a circular cross section. The Fourier coefficients DEL0, DEL1, DEL2 AND DEL3 have an obvious geometric meaning. DEL0 represents a bump, DEL1 a helical excursion, DEL2 a turning ellipse, and DEL3 a turning triangle. DEL10, DEL20 and DEL30 describe corresponding axially symmetric deformations, whereas DEL22 and DEL33 are perturbations whose toroidal periods are a half and a third of those associated with DEL2 and DEL3 respectively. Not only R and Z are coded, but their first partial derivatives are written out analytically and coded in subroutine SURF, too, in order to improve the accuracy of the calculation.

Three other wall representations appear in subroutine SURF. They have been coded and can be invoked by setting NGEOM equal to a corresponding integer. For the wall just described NGEOM = 1. For NGEOM = 2 we have the same equations but with

$$RAD = [1 + 3\ DEL3\ COS(3\overline{U} - V)]^{-1/3}.$$

For NGEOM = 3 we obtain the Heliac wall of Section 9.2, which is given by

$$R(\overline{U},V) = RBAR\ COS(TT1) + RAD\ COS(T1),$$

$$Z(\overline{U},V) = RBAR\ SIN(TT1) + RAD\ SIN(T1),$$

where

$$TT1 = -V, \qquad T1 = TT1 + DELC\ SIN(\overline{U})$$

and

$$RBAR = DELA, \qquad RAD = DELB[1 + DEL2\ COS(\overline{U})].$$

When NGEOM = 4 we come to the Wendelstein W VII-AS description of the wall, for which we refer the reader to the listing of subroutine SURF. Other wall formulas can be coded and added in subroutine SURF, provided the necessary derivatives are coded, too.

3. Initialization

The initial distribution of pressure and rotational transform per unit length are expressed in subroutine ASIN by the formulas

$$PRES = P0 \left[1 - ZPR \{ SL2(I) \}^{YPR} \right]^{XPR},$$

$$AMU = [AMU0 + AMU1\ SL1(I) + AMU2\ SL2(I)] / ZLE.$$

The mass density function is initialized from the pressure distribution. It remains fixed throughout the iteration. The pressure distribution does not remain fixed, but if it is appropriately chosen the final pressure will be reasonably close to the initial values.

The rotational transform AMU is the ratio of the derivatives of the poloidal and toroidal fluxes. One of these functions can be chosen arbitrarily with the other determined by AMU. It is best for the initial values to be close to an equilibrium solution. Therefore we start from the solution for a straight circular cylinder. In that case the solution depends on the radial coordinate alone, and the toroidal and poloidal components of the magnetic field are determined by the fluxes. The toroidal field is normalized to be near unity at $R = 1$, and the initial pressure and rotational transform distributions are solutions of the equilibrium equation for a screw pinch.

4. Grids

The parameter ALF is used to define the mesh distribution in the flux coordinate s. The flux surfaces with ALF for scaling are given by

$$R = RA(K) + SL1(I)\ RO(I,J,K)[R(J,K) - RA(K)],$$

$$Z = ZA(K) + SL1(I)\ RO(I,J,K)[Z(J,K) - ZA(K)],$$

where

$$SL1(I) = s^{ALF1}, \qquad SL2(I) = [SL1(I)]^2,$$

$$ALF1 = (1 + ALF)/2.$$

SL1(I) and SL2(I) are quantities which scale like radius and radius squared. ALF = 1 corresponds to equally spaced mesh points in a radial coordinate, whereas ALF = 0 corresponds to mesh points distributed equally in the square of that coordinate, which varies like the toroidal flux.

The solution is computed at NI, NJ and NK mesh points in the s, u and v directions. $I = 1$ corresponds to the magnetic axis and $I = $ NI corresponds to the plasma boundary. The mesh increments are $1/(NI-1)$, $1/NJ$, and $1/NK$. Because of periodicity conditions the meshes in u and v consist of $NJ+2$ and $NK+2$ points respectively. $J = 2$ corresponds to $u = 0$, $J = NJ+2$ corresponds to $u = 1$, and both represent the same point in physical space. All the calculations are done for $J = 2,3,\ldots,NJ+1$. The values at $J = 1$ and $J = NJ+2$ are defined by periodicity conditions. The same procedure is followed in the v direction.

When prescribing NI, NJ, and NK the corresponding dimensions in the code must be NI, $NJ+2$, and $NK+2$. Most of the grids we have used to study equilibrium and stability have had integral multiples of $NI \times NJ \times NK = 7 \times 12 \times 12$ mesh cells. For some cases it is necessary to refine the grid in u and v further. For example, for the Heliac we have used $6 \times 18 \times 24$, $7 \times 21 \times 28$, $8 \times 24 \times 32$, $10 \times 30 \times 40$ and $12 \times 36 \times 48$ cells in order to obtain adequate resolution. Our experience has been that refinement in the s direction alone does not improve the resolution perceptibly. It should be noted that when a stability computation is done for NRUN field periods, the dimension in v is multiplied by NRUN. Thus if the stability is done for a Heliac case with 6 field periods the arrays to carry out the analysis are multiples of $6 \times 18 \times 144$. The corresponding dimension statements should be edited to economize on core storage.

5. Convergence

The first step in the computation is the solution of an axially symmetric problem, which is used to initialize the three-dimensional calculation. When the largest of all the residuals becomes smaller than the input parameter ASYE, the two-dimensional computation terminates. In cases where the equilibrium is axisymmetric, ASYE may be set to a small value such as 0.0001 and after convergence the three-dimensional program may be used to test stability. For three-dimensional equilibrium problems, ASYE is usually set to 100 so only one iteration will be performed using the axially symmetric routines. Then the code proceeds to the three-dimensional calculation. ERR is the corresponding criterion for convergence in the solution of the three-dimensional equations. However, the calculation is usually terminated by the iteration count ITERF or by the time limit TLIM.

The input data contains the values SA1, SA2, and SA3 of the coefficients of the second time derivatives in the acceleration scheme for R, ψ, r_0 and z_0. Increasing them stabilizes but also slows down the computation. After the acceleration coefficients are chosen, the artificial time step DT must be adjusted to satisfy the Courant–Friedrichs–Lewy stability condition. The magnitude of DT depends on the problem to be solved. For the crudest grids $0.02 \leq DT \leq 0.03$ usually works. We scale DT like the mesh size divided by the square root of the Jacobian ratio. The simplest way to determine the artificial time step is to make short trial runs with a crude mesh. If DT is too large the iterations will diverge rapidly and the calculations will be terminated by a negative Jacobian. If DT is marginal the energy may oscillate so that large values of the descent coefficient occur. In either case DT should be decreased until the energy inequality DELENER < 0 is recovered. This eliminates the difficulty of an instability being masked by large descent coefficients, which ought to diminish as the acceleration scheme takes over. The initial value for the descent coefficient SE1 is read in as a large value such as 300. It remains fixed until the acceleration scheme is turned on after NAC iterations.

The parameter SEAX controls the iteration for the R equation at the magnetic axis. IF SEAX < 0 a first order equation in artificial time is used with ABS(SEAX) the coefficient of the time derivative. If SEAX > 0 a second order accelerated equation is used for the iteration with ABS(SEAX) the coefficient of the second order time derivative. There is a proportionality factor SAFI between the descent coefficients and a time average taken over NE time steps. We choose NE $= 50$ in most cases.

The parameter SEMU must also be prescribed. It is a coefficient in the rotational transform iteration, used to achieve zero net current. The quantity NVAC is an indicator which both determines whether a vacuum region occurs and how the rotational transform is treated in the equilibrium computations. IF ABS(NVAC) < 10, the rotational transform AMU is fixed. If ABS(NVAC) > 10, AMU is determined iteratively so that the net toroidal current is zero at every flux surface. The sign of NVAC determines whether there is a vacuum region. The inequality NVAC > 0 indicates a free boundary surrounded by a vacuum, and NVAC < 0 indicates a plasma region with fixed boundary.

6. Stability

For stability a second minimization is performed to calculate a perturbed energy from which the equilibrium energy is subtracted to arrive at a growth rate. The integers NR and NZ indicate Fourier coefficients of the coordinates of the magnetic axis that are to be fixed. The sign of NT determines whether a run is just for equilibrium or is also for stability. If NT > 0 only an equilibrium run is made and the selected Fourier coefficients are held fixed for the first NT iterations. Then the magnetic axis is released and the energy is minimized with no additional constraints. For a typical equilibrium run NT $= 50$.

If NT < 0 the equilibrium computation is followed by a stability analysis in which the energy is minimized subject to an additional constraint. The absolute value ABS(NT) is the number of iterations for the equilibrium calculation and must be large enough so that the equilibrium has converged to a desired accuracy. More specifically, in most runs we iterate until $-$DELENER becomes smaller than 10^{-9}. After the equilibrium energy has been computed a test function or displacement

$$DISO = (DELRAD, DELPSI, DELRO, DELZO)$$

is prescribed. The energy is minimized a second time subject to the linear constraint

$$(DIS, DISO) = DPSI,$$

where DPSI is a given amplitude and DIS is the difference between the perturbed solution and the equilibrium solution (cf. Chapter 4).

The choice of the test function DISO is made in subroutine SURF as follows:

$$DELRAD = EAM1\ EF1(J,K) + EAM2\ EF2(J,K) + EAM3\ EF3(J,K),$$

$$DELPSI = EAM4\ EG1(J,K) + EAM5\ EG2(J,K) + EAM6\ EG3(J,K),$$

$$DELRO = EL1\ EH1(K) + EL2\ EH2(K),$$

$$DELZO = EM1\ EJ1(K) + EM2\ EJ2(K).$$

For an $m = 1$ mode we put

$$EAM1(I) = 2\ [1 - SL2(I)],$$

$$EAM4(I) = -1[1 - 3\ SL2(I)],$$

$$EF1(J,K) = COS(U - V), \qquad EG1(J,K) = SIN(U - V),$$

$$EL1(K) = 1, \qquad EM1(K) = 1,$$

$$EH1(K) = COS(V), \qquad EJ1(K) = SIN(V),$$

with all the other quantities set equal to zero. For an $m = 2$ mode we put

$$EAM2(I) = 2\ [1 - SL2(I)],$$

$$EAM5(I) = -1\ [1 - 2\ SL2(I)],$$

$$EF2(J,K) = COS(2U - V), \qquad EG2(J,K) = SIN(2U - V),$$

with all the other quantities set equal to zero. For an $m = 3$ mode we put

$$EAM3(I) = 2\ SL1(I)\ [1 - SL2(I)],$$

$$EAM6(I) = -SL1(I)\ [1 - 2\ SL2(I)],$$

$$EF3(J,K) = COS(3U - V), \qquad EG3(J,K) = SIN(3U - V),$$

with all the other quantities set equal to zero, and so forth.

The equilibrium calculation is done over one field period of a configuration, but for the stability analysis the computation can be extended to several periods. Hence the stability analysis is possible for modes with larger wave lengths. The input test function for stability must take into account the fact that $0 \leq v \leq 1$ for all NRUN field periods instead of just one period, as with equilibrium. The inhomogeneous term DPSI determines the magnitude of the projection of the perturbation onto a linear subspace defined by the test function. The stability calculation is iterated to an accuracy comparable to that of the equilibrium.

7. Iteration

After NAC iterations our acceleration scheme is turned on. The integer IC specifies the frequency with which 8 out of 50 quantities selected in subroutine PRNT and used to monitor the calculation are printed, ITERF is the sum total of equilibrium plus stability iterations, and TLIM is the CP time limit after which the iterations terminate. After the time limit is reached the final values to be output are computed and plots are made; then the program stops.

8. Vacuum Region

Additional parameters must be prescribed for the vacuum computation, which is implemented by setting NVAC > 0. For this two additional lines of information are to be provided. Here NIV is the number of mesh points in the s direction for the vacuum, NV is the number of iterations for the vacuum solution per free boundary iteration, NP is the number of plasma iterations per free boundary iteration, and OM is the relaxation factor for the successive overrelaxation scheme used to solve Laplace's equation in the vacuum region. For the crudest grids OM has been set equal to 1.6 in our calculations, and for finer meshes it should be scaled with mesh size H like

$$OM = 2/(1 + H \text{ const.}).$$

The coefficient SE4 of the first order time derivative in the free boundary equation should be chosen to satisfy a Courant–Friedrichs–Lewy condition, so it scales like DT. The values for the toroidal current and poloidal current are prescribed by giving C1 and C2. For stellarators the net toroidal current is made to vanish by putting C1 = 0. The poloidal current C2 should be put equal to the length of one field period, which is $2\pi/(\text{EP QLZ})$ for toroidal equilibria and QLZ for straight equilibria.

The solution in the vacuum region is represented by a single scalar potential PT whose periods are the prescribed toroidal and poloidal currents C1 and C2. On the outer control surface, which is not a flux surface, boundary values of PT are given to model the effect of helical coils. The boundary values are specified by the relation

$$PT = C1 \ U + C2 \ V + VERT \ Z + F(W),$$

where

$$F(W) = (AMPH/XR) \ ATAN \frac{XR \ SIN(W)}{1 - XR \ COS(W)}.$$

The quantity

$$W = KW1 \ U - KW2 \ V + TORS \ SIN(AK3 \ U) - 2\pi \ USO$$

defines the winding law for a torsatron with modulation TORS.

The computation is carried out with a fixed vertical field VERT if the input parameter EVERT < 0. If EVERT > 0 the vertical field is iterated so that the average shift of the plasma surface can be prescribed by requiring it to remain fixed throughout the iteration. In this case EVERT controls the speed of the iteration.

The code has been written and vectorized to run on the CRAY computing machines of the Magnetic Fusion Energy Computing Center at the Lawrence Livermore Laboratory. For a grid with $6 \times 12 \times 12$ mesh cells an equilibrium run of 5000 time cycles requires an execution time of about 120 seconds.

9. Printed Output

Each run begins with the reading of initial data as described above. This data is immediately printed so that a documented record of the parameters of the run will be available (cf. Chapter 13). Then the axially symmetric calculation is carried out. Printout from this computation consists of the Jacobian ratio, the increment of the energy and residuals of the iteration process. Only one iteration is to be performed for initialization if we are studying a three-dimensional equilibrium. The two-dimensional results are then printed. They include the magnetic axis position, the plasma energy, and the pressure distribution. For a vacuum case (NVAC > 0) the origin of the coordinate system, vacuum energy, Hamiltonian and Fourier coefficients for the free boundary function are printed.

The three-dimensional equilibrium computation comes next. The residuals of the computation in their dependence on the iteration frequency IC follow. The quantities printed are those which are selected in the input file on Card 26. After the last iteration the final values of the energy and Fourier coefficients of the magnetic axis are displayed. If NVAC > 0 the Fourier coefficients of the vacuum axis and the free boundary are also printed. For the plasma region the average on each flux surface of physical quantities such as magnetic fields, currents, rotational transform, pressure, plasma β, and p' are given as functions of the average radius. For the vacuum region the toroidal current and poloidal current and the vertical field are presented. The average β and the well depth are computed and printed. The island width and Jacobian values, their maxima, minima, and various norms appear as functions of the flux. Then a page of the Fourier coefficients of the equilibrium with respect to flux coordinates is printed. The Mercier criterion has been coded and is included. The individual terms contributing to the Mercier criterion, the Mercier criterion itself, the well depth V''/V' and the mirror ratio T are tabulated as functions of the toroidal flux. Two different evaluations of the current or island width term Ω_λ or Ω_d are given. After that appear Fourier coefficients of the eigenfunction calculated in a stability run.

Finally a time history of the Fourier coefficients of the magnetic axis and the Fourier coefficients of a flux surface at some fixed radius are presented. Column headings are the poloidal and toroidal mode numbers m and k. The Fourier coefficients of the flux surface refer to RO(I,J,K). The value of I can be chosen by setting the indicator I1 in the main routine. For a vacuum run the time history of the Fourier coefficients of the free boundary is printed. It is followed by that of the descent coefficient.

10. Plots

Some of the results of the computation are presented graphically using a graphics library; TV80LIB, available on the CRAY machine at Livermore. However, the coding for the plots in subroutine TPLOT can be adapted to any plot package.

The first plot gives the flux surfaces at four different equally spaced values of the toroidal coordinate v. The next page consists of the pressure and rotational transform plotted as functions of the radius, together with the Mercier criterion and island width in their dependence on the toroidal flux. Then the energy and R, ψ, r_0 and z_0 residuals for the plasma are plotted as functions of artificial time. A plot of residuals of the axis distortion and the rotational transform in their dependence on artificial time follows. The time history of the magnetic axis and of the Fourier coefficients of the radius are graphed. Then the poloidal and toroidal fields and currents appear. For the vacuum case, free boundary, axis and vacuum residuals as well as the Fourier coefficients of the free boundary are plotted.

CHAPTER 12
Glossary

Most of the input data appear in a file (TAPE 25) which is made up of pairs of input cards or lines of text. The first line of the pair is a dummy which names the parameters. The second line gives the numerical values of these parameters in corresponding locations. Each parameter occupies 8 columns of its card or line.

CARDS 1,3,5,...   FORTRAN names of the parameters.

CARD 2

    EP

The inverse ε of the major radius or aspect ratio A; $0 \leq EP < 1$. FORMAT F8.4.

    RBOU

The ratio of the plasma radius to the wall radius. It is equal to 1 if there is no vacuum region. FORMAT F8.4.

    QLZ

The number Q of field periods if EP > 0. If EP = 0, it is the cylinder length. FORMAT F8.4.

    NRUN

The number n of field periods for which a stability calculation is performed. NRUN = 1 for equilibrium runs. FORMAT I8.

    NGEOM

This selects the appropriate wall geometry. NGEOM = 1 selects the standard wall for a device like ATF-1. NGEOM = 2 selects a similar wall description.

NGEOM = 3 selects the wall for the Heliac and NGEOM = 4 selects a more complicated wall for the Wendelstein W VII-AS. FORMAT I8.

RUN   The run number. FORMAT I8.

CARD 4

DEL0, DEL1, DEL2, DEL3, DEL10, DEL20, DEL30

The Fourier coefficients in the outer wall equations which appear in subroutine SURF. Other wall formulas may also be prescribed. FORMAT 7F8.4.

CARD 6

DEL22, DEL33, DELA, DELB, DELC

Additional coefficients for the wall. FORMAT 5F8.4.

CARD 8

XPR, YPR, ZPR

Parameters used in the definition of the pressure $PRES = PO[1 - ZPR(SL2)^{YPR}]^{XPR}$. $XPR = 2$, $YPR = 1$ and $ZPR = 1$ in most of the runs. FORMAT 3F8.4.

P_0 The maximum pressure p_0. FORMAT F8.4.

AMU0, AMU1, AMU2

The coefficients appearing in the initial expression for the rotational transform μ. FORMAT 3F8.4.

CARD 10

ALF

A scaling constant which defines the mesh distribution in the radial direction. ALF = 0 means the radial coordinate is proportional to the toroidal flux. ALF = 1 means the radial coordinate is proportional to the radius. Suggested value ALF = 0.5. FORMAT F8.4.

ALFU

A zoning parameter which is used to redistribute the mesh in u so that it becomes as equally spaced as possible. Suggested value 0 to avoid rezoning. For the Heliac we use ALFU = −0.5. FORMAT F8.4.

CARD 12

NI	The number of mesh points in the s direction for the plasma region. FORMAT I8.
NJ	The number of mesh points in the u direction. FORMAT I8.
NK	The number of mesh points in the v direction. FORMAT I8.
ASYE	The convergence threshold for the axisymmetric part of the calculation. Suggested value for a two-dimensional calculation $1.0E-04$. For a three-dimensional calculation ASYE $=1.0E+02$. FORMAT E8.1.
ERR	A convergence threshold for the three-dimensional calculation. Suggested value $1.0E-18$. FORMAT E8.1.

CARD 14

SA1	The coefficient of the second time derivative of R. Suggested value 4.0. FORMAT F8.4.
SA2	The coefficient of the second time derivative of ψ. Suggested value 4.0. FORMAT F8.4.
SA3	The coefficient of the second time derivative of the magnetic axis. Suggested value 0.2. FORMAT F8.4.
DT	The artificial time step. For the crudest grids usually $.02 \leq DT \leq .03$. For finer meshes DT scales like mesh size divided by the square root of the Jacobian ratio. FORMAT F8.4.

CARD 16

SE1	The initial value of the descent coefficient in the iterative scheme for the plasma equations. Suggested value 300.0. FORMAT F8.4.
SEMU	The descent coefficient for the rotational transform iteration used to achieve zero net current. The typical range for this parameter is $0.1 \leq SEMU \leq 1.0$. Suggested value 0.2. FORMAT F8.4.

SEAX   This controls the iteration for the R equation on the magnetic axis. If SEAX < 0 the iteration uses a first order time equation with coefficient ABS(SEAX). If SEAX > 0 the iteration uses a second order time equation and ABS(SEAX) is the coefficient of the second derivative. For most cases $1 \le$ SEAX ≤ 2 is adequate. If there are convergence problems for the R equation at the axis, SEAX ≤ -4 should be tried. This slows the computation but usually improves convergence in difficult cases. FORMAT F8.4.

CARD 18

SAFI   A proportionality constant for the acceleration scheme in the plasma region. Suggested value 2.2. FORMAT F8.4.

NE   The number of points used in computing an average for the descent coefficient. NE ≤ 100. FORMAT I8.

NVAC   An indicator which specifies the type of case to be computed. If NVAC > 0 a vacuum region is included in the calculation. If NVAC < 0 only a fixed boundary plasma is being studied. If ABS(NVAC) < 10 the rotational transform is held constant. If ABS(NVAC) > 10 the rotational transform is determined by imposing the condition of zero net current. FORMAT I8.

CARD 20

NR, NZ   Indices of the magnetic axis Fourier coefficients which are to be fixed in the calculation. $1 \le$ NR ≤ 7, $1 \le$ NZ ≤ 7. FORMAT 2I8.

NT   An indicator which determines the kind of run to be made. If NT > 0 an equilibrium calculation will be made; suggested values $50 \le$ NT ≤ 100. If NT < 0 a stability calculation will be made; after ABS(NT) iterations the equilibrium is extended to NRUN field periods and a stability analysis is carried out. FORMAT I8.

NAC   An iteration count after which the acceleration scheme is invoked. Suggested value NAC $= 200$. FORMAT I8.

DPSI	For a stability run this determines the magnitude of the projection of the perturbed solution onto a linear subspace defined by a given test function. Generally DPSI = 0.2 for $m = 1$ modes and DPSI = 0.05 for $m = 2$ modes. FORMAT F8.4.
GAM	The adiabatic gas constant γ. FORMAT F8.4.
IGAM	This determines the density used in the norm for the stability problem. If IGAM < 0 that density is constant, but if IGAM > 0 the usual equation of state prevails. FORMAT I8.

CARD 22

MIS, NIS	Poloidal and toroidal mode numbers for which Fourier coefficients of $J \cdot B / B^2$ are printed. FORMAT 2I8.
NAS	This controls the initial values for R. If NAS > 0 they are equal to the axisymmetric values. If NAS < 0 they are computed using a better approximation for the three-dimensional case. FORMAT I8.
IROT	This is used for output only. The rotational transform is shifted by IROT. FORMAT I8.
MF, NF, IPF	The indices MF and NF determine how many terms are kept in the Fourier analysis of $1/B^2$, and Fourier coefficients are printed at $I = $ IPF. Suggested values MF = 6 and NF = 3. FORMAT 3I8.

CARD 24

IC	After every IC time cycles the code prints the iteration residuals during the computation. For production runs IC = 50 is suggested. FORMAT I8.
ITERF	This sets the total number of iterations for equilibrium and stability that are to be calculated. FORMAT I8.
TLIM	The CP time limit in seconds. It is usually made slightly larger than the expected time for the run so that the run will be terminated by the number ITERF of iterations. FORMAT F8.1.

CARD 26

PRINT1,
PRINT2,...,
PRINT8

Eight out of 50 parameters which can be printed every IC iterations by insertion of the appropriate names on this card. The 50 parameters appear in subroutine PRNT. They are used to monitor the solution. FORMAT 8A8.

CARD 28

USO

Phase in winding law. FORMAT F8.4.

UIN

This shifts the poloidal harmonics in the Fourier analysis. The coefficients are computed for arguments $U + UIN$, $2U + UIN$, $3U + UIN$; usually UIN $= 0$. FORMAT F8.4.

USM

This determines the boundary condition on the potential at the outer wall for a vacuum run. For USM < 0 a Dirichlet condition is applied to the entire boundary. For USM > 0 a Dirichlet condition is used in the interval $USM \leq U \leq 1 - USM$ and a Neumann condition is used on the remainder of the boundary. Suggested value USM $= -1$. FORMAT F8.4.

IVAX

This determines the treatment of the vacuum axis. If IVAX < 0 the vacuum axis is not iterated. If IVAX > 0 the vacuum axis is iterated. FORMAT I8.

CARD 30

NIV

The number of mesh points in the s direction in the vacuum region. NIV ≥ 6 for NVAC > 0 and NIV $= 1$ for NVAC < 0. FORMAT I8.

NV

The number of vacuum iterations per free boundary iteration. NV ≥ 2 is required, and NV $= 3$ is suggested. FORMAT I8.

NP

The number of plasma iterations per free boundary iteration. NP $= 1$ is suggested. FORMAT I8.

OM	The relaxation factor for Laplace's equation in the vacuum region. For the crudest grids OM = 1.6 is suggested, and OM should be scaled with mesh size H like $2/(1 + H$ const.$)$. FORMAT F8.4.
SAFV	A positive proportionality constant for the acceleration scheme in the vacuum. Suggested value 2.0. FORMAT F8.4.
SE4	The descent coefficient for the free boundary equation. SE4 > 0. Suggested value 1.0. FORMAT F8.4.
C1	The toroidal current defined by the scalar potential in the vacuum region. C1 = 0 for stellarators. FORMAT F8.4.
C2	The poloidal current defined by the scalar potential in the vacuum region. Recommended value $2\pi/($EP QLZ$)$ if EP > 0 and QLZ if EP = 0. FORMAT F8.4.

CARD 32

KW1, KW2, AK3, TORS	These parameters specify the winding law. FORMAT 2I8, 2F8.4.
AMPH	This controls an amplitude in the winding law. FORMAT F8.4.
EVERT	This determines the treatment of the vertical field. If EVERT < 0 the vertical field is held fixed. If EVERT > 0 the vertical field is iterated, using EVERT, to produce a fixed shift; suggested value 1.0. FORMAT F8.4.
WRAD	A smoothing factor WRAD = 1/XR in the winding law. Suggested value 1.7. FORMAT F8.4.
VERT	The vertical field strength. Suggested values $-.01 \leq$ VERT $\leq .01$. FORMAT F8.4.

CARD 34

NRA1, NRA2, NZA1, NZA2,	These parameters are used to choose quantities which are to be plotted at the end of the computation.

MK1, MK2, MK3, MK4	There are seven possible choices of NRA and NZA which specify Fourier coefficients of the magnetic axis. See subroutine FPRINT. The MK are indices specifying Fourier coefficients of R. FORMAT 8I8.

CARD 36

EL1, EL2	Coefficients of DELRO in the test function for stability. FORMAT 2F8.4.
EM1, EM2	Coefficients of DELZO in the test function for stability. FORMAT 2F8.4.
M,N	Mode numbers of the test function for stability. FORMAT 2I8.

CARD 38

EAM1, EAM2, EAM3	Coefficients of DELRAD in the test function for stability. FORMAT 3F8.4.
EAM4, EAM5, EAM6	Coefficients of DELPSI in the test function for stability. FORMAT 3F8.4.

CHAPTER 13
Sample Run

INPUT DATA AS READ FROM INPUT DATA DECK

EP	RBOU	QLZ	NRUN	NGEOM	RUN
0.1000	1.0000	18.0000	1	1	1001

DEL0	DEL1	DEL2	DEL3	DEL10	DEL20	DEL30
0.0000	0.0000	0.2700	0.0000	0.0000	0.0000	0.0000

DEL22	DEL33	DELA	DELB	DELC
0.0000	0.0000	0.0000	0.0000	0.0000

XPR	YPR	ZPR	P0	AMU0	AMU1	AMU2
2.0000	1.0000	1.0000	0.0100	0.0300	0.0000	0.0300

ALF	ALFU
0.5000	0.0000

NI	NJ	NK	ASYE	ERR
7	12	12	1.0E+02	1.0E-18

SA1	SA2	SA3	DT
4.0000	4.0000	0.2000	0.0300

SE1	SEMU	SEAX
300.00	0.20	2.00

SAFI	NE	NVAC
2.2000	50	-20

NR	NZ	NT	NAC	DPSI	GAM	IGAM
1	1	100	200	0.000	2.000	1

MIS	NIS	NAS	IROT	MF	NF	IPF
2	1	-5	0	6	3	4

IC	ITERF	TLIM
50	2000	500.0

PRINT1	PRINT2	PRINT3	PRINT4	PRINT5	PRINT6	PRINT7	PRINT8
RO ERR	PSI ERR	DELENER	ENERGY				

USO	UIN	USM	IVAX
0.5000	0.0000	-1.0000	1

NIV	NV	NP	OM	SAFV	SE4	C1	C2
1	3	1	1.6000	2.0000	1.0000	0.0000	0.0000

KW1	KW2	AK3	TORS	AMPH	EVERT	WRAD	VERT
0	0	1.0000	0.0000	0.0000	-1.0000	1.7000	0.0000

NRA1	NRA2	NZA1	NZA2	MK1	MK2	MK3	MK4
1	3	1	2	11	21	32	10

EL1	EL2	EM1	EM2	M	N
0.0000	0.0000	0.0000	0.0000	0	0

EAM1	EAM2	EAM3	EAM4	EAM5	EAM6
0.0000	0.0000	0.0000	0.0000	0.0000	0.0000

3D RESULTS

ITER	RO ERR	PSI ERR	DELENER	ENERGY
1	1.0983E+00	1.3236E+00	6.5501E+00	6.55007458E+00
51	6.4874E-01	7.2254E-01	-7.6414E-04	6.49100792E+00
101	4.2178E-01	4.2867E-01	-3.3295E-04	6.46554610E+00
151	2.9903E-01	2.9784E-01	-1.9235E-04	6.45214357E+00
201	2.2873E-01	2.2198E-01	-1.0121E-04	6.44514365E+00
251	1.6482E-01	3.4314E-01	-5.3080E-04	6.44194294E+00
301	8.2119E-02	7.3733E-02	-5.2911E-05	6.43114378E+00
351	4.2842E-02	3.6252E-02	-1.9072E-05	6.42943804E+00
401	1.5996E-02	1.4786E-02	-2.9415E-06	6.42900347E+00
451	6.2579E-03	4.7532E-03	-1.1416E-06	6.42891996E+00
501	3.9569E-03	2.3523E-03	-1.1329E-06	6.42886555E+00
551	3.5698E-03	1.3206E-03	-1.5993E-06	6.42879848E+00
601	2.9799E-03	1.2701E-03	-2.0057E-06	6.42870575E+00
651	2.2443E-03	1.0907E-03	-1.5871E-06	6.42861291E+00
701	1.7275E-03	9.4752E-04	-8.7893E-07	6.42855218E+00
751	1.2814E-03	9.1654E-04	-4.3470E-07	6.42852086E+00
801	9.1282E-04	8.7707E-04	-2.3927E-07	6.42850484E+00
851	7.1252E-04	8.2140E-04	-1.6109E-07	6.42849515E+00
901	6.5001E-04	7.4599E-04	-1.2523E-07	6.42848811E+00
951	5.7989E-04	6.5327E-04	-1.0366E-07	6.42848243E+00
1001	5.0137E-04	5.6271E-04	-8.5276E-08	6.42847772E+00
1051	4.1444E-04	4.7746E-04	-6.7058E-08	6.42847392E+00
1101	3.2204E-04	4.0432E-04	-5.0146E-08	6.42847101E+00
1151	2.6945E-04	3.3949E-04	-3.6364E-08	6.42846887E+00
1201	2.3460E-04	2.8276E-04	-2.6201E-08	6.42846732E+00
1251	2.0164E-04	2.3687E-04	-1.8954E-08	6.42846621E+00
1301	1.7475E-04	1.9932E-04	-1.3739E-08	6.42846540E+00
1351	1.4788E-04	1.6881E-04	-9.9230E-09	6.42846481E+00
1401	1.2242E-04	1.4334E-04	-7.1371E-09	6.42846439E+00
1451	1.0085E-04	1.2138E-04	-5.1345E-09	6.42846409E+00
1501	8.3543E-05	1.0296E-04	-3.7178E-09	6.42846387E+00
1551	7.0386E-05	8.7197E-05	-2.7173E-09	6.42846371E+00
1601	5.8763E-05	7.3752E-05	-2.0048E-09	6.42846359E+00
1651	4.9074E-05	6.2211E-05	-1.4879E-09	6.42846351E+00
1701	4.1079E-05	5.2277E-05	-1.1105E-09	6.42846344E+00
1751	3.4933E-05	4.3869E-05	-8.3207E-10	6.42846340E+00
1801	2.9991E-05	3.6826E-05	-6.2747E-10	6.42846336E+00
1851	2.5630E-05	3.0892E-05	-4.7555E-10	6.42846333E+00
1901	2.1743E-05	2.5915E-05	-3.6277E-10	6.42846331E+00
1951	1.8343E-05	2.1719E-05	-2.7816E-10	6.42846330E+00
2001	1.5388E-05	1.8222E-05	-2.1473E-10	6.42846328E+00

EXECUTION TIME=   49.46

PLASMA ENERGY= 6.428463284E+00

FOURIER COEFFICIENTS OF MAGNETIC AXIS

	CONST	SIN(V)	COS(V)	SIN(2V)	COS(2V)
R	0.01085	0.00004	-0.00108	-0.00000	0.00000
Z	-0.00026	-0.00108	-0.00004	0.00000	0.00000

FOURIER COEFFICIENTS OF FLUX SURFACE AT I= 4, S= 0.35

	CONST	SIN(V)	COS(V)	SIN(2V)	COS(2V)
CONST	0.96916	0.00000	0.00018	-0.00000	0.00000
SIN(U)	-0.00012	0.00094	-0.00000	-0.00000	-0.00000
COS(U)	0.00589	0.00000	0.00092	-0.00000	-0.00000
SIN(2U)	-0.00004	0.07476	0.00000	-0.00005	0.00000
COS(2U)	0.00001	-0.00000	0.07476	-0.00000	-0.00005
SIN(3U)	-0.00000	-0.00139	-0.00001	0.00012	-0.00000
COS(3U)	-0.00003	0.00001	-0.00139	0.00000	0.00012

FLUX SURFACE AVERAGE PLASMA PARAMETERS PER FIELD PERIOD

RADIUS	BTOR	BPOL	TCURR	PCURR	MU	P	BETA
0.000	1.152	0.000	0.000	4.025	0.040	0.013	0.020
0.251	1.151	0.017	-0.000	4.045	0.040	0.012	0.017
0.423	1.145	0.027	-0.000	4.064	0.040	0.009	0.013
0.576	1.133	0.036	-0.000	4.075	0.046	0.005	0.008
0.720	1.115	0.043	0.000	4.082	0.056	0.002	0.004
0.860	1.090	0.047	0.000	4.083	0.070	0.001	0.001
1.000	1.069	0.046	0.000	4.138	0.087	0.000	0.000

AVERAGE BETA= 0.007   WELL DEPTH= -0.192   IROT= 0

	J.B/B.B			JACOBIAN	
FLUX	NORM	MAX	MIN	MAX JAC	MIN JAC
0.00				3.050	2.377
0.07	0.032	0.237	-0.136	3.178	2.325
0.19	0.030	0.216	-0.113	3.158	2.375
0.35	0.025	0.178	-0.109	3.109	2.552
0.54	0.023	0.156	-0.122	2.980	2.846
0.76	0.037	0.219	-0.259	3.402	2.680
1.00				4.037	2.508

J.B/B.B FROM CURL

FLUX	CONST	COS(MU-NV)	SIN(MU-NV)
0.07	-0.0084	0.1674	-0.0000
0.19	-0.0113	0.1591	-0.0000
0.35	-0.0100	0.1312	-0.0000
0.54	-0.0099	0.1230	0.0000
0.76	-0.0239	0.2015	0.0000

ISLAND WIDTH NORMS

D(J.B/B.B)= 0.02564   D(COS21)= 0.02404   D(SIN21)= 0.00000

FOURIER COEFFICIENTS OF J.B/B.B.P′ AT S= 0.35 VERSUS PSI AND PHI

	CONST	SIN(PHI)	COS(PHI)	SIN(2PHI)	COS(2PHI)
CONST	0.00000	-0.00000	-0.00000	0.00000	-0.00000
SIN(PSI)	0.00017	-0.00764	-0.00007	0.00001	0.00000
COS(PSI)	-2.37526	-0.00007	-0.00770	0.00000	0.00001
SIN(2PSI)	0.00052	-0.25987	-0.00011	-0.00006	-0.00000
COS(2PSI)	-0.11233	-0.00011	-0.25988	-0.00000	-0.00006
SIN(3PSI)	0.00011	-0.04777	-0.00011	-0.00390	-0.00000
COS(3PSI)	-0.00469	-0.00011	-0.04778	-0.00000	-0.00390

FOURIER COEFFICIENTS OF 1/B.B AT S= 0.35 VERSUS PSI AND PHI

	CONST	SIN(PHI)	COS(PHI)	SIN(2PHI)	COS(2PHI)
CONST	0.76729	0.00001	0.00042	-0.00000	0.00000
SIN(PSI)	-0.00001	-0.00564	-0.00005	0.00001	-0.00000
COS(PSI)	0.08309	0.00005	-0.00560	-0.00000	0.00001
SIN(2PSI)	-0.00002	-0.09066	-0.00004	-0.00004	-0.00000
COS(2PSI)	0.00393	0.00004	-0.09065	0.00000	-0.00004
SIN(3PSI)	-0.00000	-0.01055	0.00002	-0.00186	-0.00000
COS(3PSI)	0.00016	-0.00002	-0.01055	0.00000	-0.00186

SPLINE INTERPOLATED VALUES OF MERCIER RESULTS

FLUX	MERCIER	SHEAR2	JDOTB	WELL	CURRENT
0.20	-0.06749	0.00162	0.00039	-0.04389	-0.02561
0.25	-0.05740	0.00375	0.00066	-0.04222	-0.01959
0.30	-0.04779	0.00597	0.00088	-0.04009	-0.01455
0.35	-0.03866	0.00828	0.00106	-0.03751	-0.01048
0.40	-0.03002	0.01065	0.00120	-0.03450	-0.00737
0.45	-0.02198	0.01296	0.00129	-0.03118	-0.00506
0.50	-0.01466	0.01507	0.00135	-0.02771	-0.00337
0.55	-0.00820	0.01681	0.00136	-0.02424	-0.00213
0.60	-0.00268	0.01808	0.00132	-0.02088	-0.00120
0.65	0.00189	0.01887	0.00125	-0.01765	-0.00058
0.70	0.00550	0.01917	0.00112	-0.01454	-0.00025
0.75	0.00816	0.01898	0.00096	-0.01155	-0.00022

CHAPTER 14
Listing of Code BETA with Comment Cards

```
C                        TABLE OF SUBROUTINES
C
C
C      1. BETA  - MAIN CONTROL PROGRAM                         88
C
C      2. ASYM  - INITIALIZATION, CONTROL OF 2D SOLUTION      105
C
C      3. GRAD  - AXISYMMETRIC EQUATIONS                      110
C
C      4. ASOR  - ITERATION OF 2D VACUUM EQUATIONS            113
C
C      5. ASBO  - ITERATION OF FREE BOUNDARY IN 2D            115
C
C      6. EXTEN - ADDITION OF PERIODS FOR STABILITY           117
C
C      7. PROJ  - INITIALIZATION FOR STABILITY                120
C
C      8. TGRAD - ITERATION OF 3D PLASMA EQUATIONS            122
C
C      9. CIN   - COEFFICIENTS OF PLASMA EQUATIONS            130
C
C     10. CBO   - BOUNDARY CONDITIONS                         135
C
C     11. CDEN  - CURRENT DENSITY AND MERCIER CRITERION       136
C
C     12. FPHI  - FOURIER COEFFICIENTS                        142
C
C     13. TSOR  - ITERATION OF 3D VACUUM EQUATIONS            147
C
C     14. CV    - COEFFICIENTS OF VACUUM EQUATIONS            152
C
C     15. TBO   - ITERATION OF FREE BOUNDARY IN 3D            152
C
C     16. ELAM  - ACCELERATION SCHEME                         156
C
C     17. BNORM - NORM FUNCTION                               157
C
C     18. SPLIF - GENERAL PURPOSE CUBIC SPLINE                157
C
C     19. INTPL - GENERAL PURPOSE INTERPOLATION               158
C
C     20. PINT  - POLYNOMIAL INTERPOLATION                    159
C
C     21. ASOUT - PRINTOUT OF 2D SOLUTION                     160
C
C     22. FPRINT- PRINTOUT OF 3D SOLUTION                     161
C
```

```
C    23. PRNT  - SELECTION OF PRINTED OUTPUT              170
C
C    24. TPLOT - PLOT OF 3D SOLUTION                      171
C
C    25. PLOTB - CONTROL OF PLOT PACKAGE                  181
C
C    26. PBOU  - BOUNDARY VALUES FOR VACUUM POTENTIAL     182
C
C    27. ASIN  - INITIALIZATION OF MASS AND FLUX          183
C
C    28. SURF  - WALL FORMULAS AND TEST FUNCTION          186
C
C
C
C              INDEX OF LIBRARY ROUTINES
C
C    SECOND - RETURNS CUMULATIVE CPU TIME                     99
C    FO4ATF - LINEAR EQUATION SOLVER               121,160,190
C    KEEP80 - SAVE PLOT FILE FOR LATER USE                  172
C    FR80ID - MAKE AN FR80 FILE                             172
C    MAP    - TRANSLATE AND SCALE                           172
C    LINE   - DRAW A LINE                                   175
C    SETCRT - SET INITIAL POSITION FOR PLOT         176,177,182
C    VECTOR - DRAW A LINE                           176,177,182
C    SETLCH - PLOT IN USER'S COORDINATES            176,177,182
C    CRTBCD - WRITE LABELS ON AXES                  176,177,182
C    FRAME  - ADVANCE FRAME                         177,178,179
C    GAXIS  - DRAW AND LABEL AXES                           182
C
C
C
C    COMMON BLOCKS AND DIMENSION STATEMENTS WHICH NEED TO BE CHANGED
C    BECAUSE THEY DEPEND ON THE GRID SIZE APPEAR JUST ONCE AT THE
C    BEGINNING OF THE CODE AS CLICHE STATEMENTS . TO REFINE THE GRID
C    ONLY CLICHE NAMEA IS CHANGED. THE PARAMETERS NID,NJD,NKD,NRUND,
C    AND NIVD CORRESPOND TO NI,NJ,NK,NRUN, AND NIV RESPECTIVELY.
C    NKD MUST BE .GE. NJD IN THIS STATEMENT.
C
     CLICHE NAMEA
     PARAMETER(NID=09,NJD=16,NKD=16,NRUND=01,NIVD=1)
     ENDCLICHE
     CLICHE NAMEB
     USE NAMEA
     PARAMETER(ID=NID,JD=NJD+2,KD=NKD*NRUND+2,IV=NIVD)
     ENDCLICHE
     CLICHE NAME1
     USE NAMEB
     COMMON RO(ID,JD,KD), AL(ID,JD,KD), XO(ID,JD,KD), XL(ID,JD,KD),
    1 R(JD,KD), Z(JD,KD), RU(JD,KD), ZU(JD,KD), RV(JD,KD),ZV(JD,KD)
    2 , X(JD,KD), RA(KD), ZA(KD), RN(KD), ZN(KD), RB1(KD), RB2
    3 (KD), ZB1(KD), ZB2(KD), RBU1(KD), RBU2(KD), ZBU1(KD), ZBU2(
    4 KD), RBV1(KD), RBV2(KD), ZBV1(KD), ZBV2(KD), HB1(KD),HB2(KD)
    5 , D1(KD), D2(KD), D3(KD), D4(KD), D5(KD), D6(KD), D7(KD
```

```
6 ), D8(KD), D9(KD), D10(KD), D11(KD), D12(KD), D13(KD), AKB(
7 JD,KD), RPAX(KD), RPBX(KD), RPCX(KD), ERRO1(KD), ERRO2(KD)
  ENDCLICHE
  CLICHE NAME2
  USE NAMEB
  COMMON /AUX/ RR2, ZZ2, RRA, ZZA, XX(KD), E1(ID,KD), E2(ID,KD),
1 F1(ID,KD), F2(ID,KD), G1(ID,KD), G2(ID,KD), P1(ID,KD), P2(ID
2 ,KD), Q1(ID,KD), Q2(ID,KD), V1(ID,KD), V2(ID,KD), U1(ID,KD)
3 , U2(ID,KD), UV1(ID,KD), UV2(ID,KD), D(ID,KD), DD(ID,KD), BK
4 (ID,KD), DM(ID,KD), PM1(ID,KD), PM2(ID,KD), CA1(KD), CA2(KD
5 ), CB1(KD), CB2(KD), CC1(KD), CC2(KD), CD1(KD), CE1(KD),
6 CE2(KD), CF1(KD), CF2(KD), CN2(KD), XA1(KD), XA2(KD), XB1(KD)
7 , XB2(KD), XC1(KD), XC2(KD), CG1(KD), CG2(KD), CL1(KD),
8 CL2(KD), CM1(KD), CM2(KD), CN1(KD), XG1(KD), XG2(KD), XN1(KD)
9 , XN2(KD), XD1(KD), XD2(KD), XF1(KD), XF2(KD), XP1(KD),
$ XP2(KD), XQ1(KD), XQ2(KD), CAF1(KD), CAF2(KD), CAG1(KD),
$ CAG2(KD), ROAX(KD), ROBX(KD), ROCX(KD), SUMRO(KD), SUMR(KD)
$ , SUMZ(KD), SUMA(KD), SUMB(KD), SUMC(KD), CAH1(KD), CAH2(KD
$ )
  ENDCLICHE
  CLICHE NAME3
  USE NAMEB
  COMMON /AUX/ R2, Z2, RA, ZA, X(KD), R(ID,KD), AL(ID,KD), PA(ID,
1 KD), D(ID,KD), E(ID,KD), F(ID,KD), G(ID,KD), R1(ID,KD), Z1(
2 ID,KD), GR(ID,KD), GZ(ID,KD), DD(ID,KD), QR(ID,KD), QZ(ID,KD),
3 A(KD), B(KD), C(KD), H(KD), RR(KD), ZZ(KD), RU(KD), ZU
4 (KD), RB(KD), ZB(KD), RBU(KD), ZBU(KD), XA(KD), XC(KD),
5 XPD(KD), XPK(KD), AK2(KD), AKF(KD), AKFU(KD), GAXR, GAXZ, AF
6 (KD), AG(KD), SUMA, SUMB, SUMC, AH(KD)
  ENDCLICHE
  CLICHE NAME4
  USE NAMEB
  COMMON /AUX/ VA1(ID,KD), VA2(ID,KD), VB1(ID,KD), VB2(ID,KD), VC1(I
1D,KD), VC2(ID,KD), VD1(ID,KD), VD2(ID,KD), VE1(ID,KD), VE2(ID,KD)
2 , VL1(ID,KD), VL2(ID,KD), D14(KD), D15(KD)
  ENDCLICHE
  CLICHE NAME5
  USE NAMEB
  COMMON /AUX/ AK(JD,KD),AKFU(JD,KD),AKFV(JD,KD),AKF(JD,KD)
  ENDCLICHE
  CLICHE NAME6
  USE NAMEB
  COMMON /EIGEN/ EF1(JD,KD), EF2(JD,KD), EG1(JD,KD), EG2(JD,KD),
1 EH2(KD), EJ1(KD), EJ2(KD), EAM1(ID), EAM2(ID), EAM3(ID), EAM4(
2 ID), EH1(KD), EL1, EL2, EM1, EM2, RQ, PNORM, IRQ, PRQ, ANORM,
3 EF3(JD,KD), EG3(JD,KD), EAM5(ID), EAM6(ID), EBD(JD,KD), BNORM,
4 QNORM
  ENDCLICHE
  CLICHE NAME7
  USE NAMEB
  COMMON /FOU/ SV(7,KD), SU(7,KD), SFI(7,KD), SRO(25,25), XR(7), XZ
1 (7), YR(7), SUP(7,KD), SVP(7,KD), YZ(7), OP1(200), OP2(200),
2 OP3(100), STO(7,7), SNORM, DELTE, EVAL, NT
```

```
ENDCLICHE
CLICHE NAME8
USE NAMEB
COMMON /PLOT/ NRA1, NRA2, NZA1, NZA2, MK1, MK2, MK3, MK4, M1, M2,
1M3,M4, K1, K2, K3, K4, RNAME(7), ZNAME(7), RP(ID), RM(ID), RP1(ID)
2 , RV1(ID)
ENDCLICHE
CLICHE NAME9
USE NAMEB
COMMON /POT/ RVA(KD), ZVA(KD), BPV(KD), BTV(KD), PT(IV,JD,KD)
ENDCLICHE
CLICHE NAME10
USE NAMEB
COMMON /INP/ Q(ID), QT(ID), AM(ID), PR(ID), QQ(ID), PC(ID), TC(ID)
1 , NGEOM, BPP(ID), BTP(ID), BET(ID), SPR(ID), SPRM(ID), BMP(ID),
2 BMM(ID), SL1(ID), SL2(ID), SA(ID), PS(ID), QS(ID), PQS(ID), XS(ID
3 ), DP(ID), DQ(ID), SPM(ID), PRP(ID), PRM(ID), VOL(ID), SPVOL(ID),
4 SPPVOL(ID), AMS(ID), MIS1 , MIS, NIS, NAS,MF,NF,IPF,AMAGM
ENDCLICHE
CLICHE NAME11
COMMON /AC/ EN(100), OP(100), OE(100), ET(5), AA1(5), AA2(5), AA3(
1 5), AVE(5), NAC, NE, SAFI, SAFPSI, SAFRO, SAFAX
ENDCLICHE
CLICHE NAME12
USE NAMEB
COMMON /CPL/ NI, NJ, NK, EP, ZLE, GAM, SM, N1, N2, N3, N4, N5,
1 NVAC, HS, HU, HV, PI, RX, RY, E4, A1, A2, A3, A4, A5, A6, HU4,
2 HV4, HUV, IC, IO, SA1, SA2, SA3, SE1, SE2, SE3, DT, RA1, RA2, RA3
3 , RE1, RE2, RE3, ENER, FAXIS, DG1, DG2, DG3, DH1, DH2, DH3, NPLOT
4 , I1, ROA, ROB, ROC, IROT, SNL2(ID)
ENDCLICHE
CLICHE NAME13
COMMON /CVA/ NIV, FV1, FV2, NV, NP, OM, PM, HR, H1, H2, H3, C1, C2
1 , N6, EVAC, ETOT, A11, A22, A12, HR4, DTB, FAXV, OM1, OM2, P01,
2 P02, SEAX, USM, IVAX
ENDCLICHE
CLICHE NAME14
COMMON /FUNC/ ALF, RBOU, DEL0, DEL1, DEL2, DEL3, DEL10, DEL20,
1 DEL30, DEL22, DEL33, P0, XPR, AMU0, AMU1, AMU2, AMP,NRUN,ALFU
2,DELA,DELB,DELC,YPR,ZPR
ENDCLICHE
CLICHE NAME15
USE NAMEB
COMMON /WIND/ ITOR, KW1, KW2, AK3, TORS, AMPH, WRAD, VERT, WRADV,
1 ZVERT, US0, RCOIL(KD,5), ZCOIL(KD,5)
ENDCLICHE
CLICHE NAME16
COMMON /GABO/ DEL1R, DEL1Z, DEL(3,4)
ENDCLICHE
CLICHE NAME17
USE NAMEB
COMMON RO(ID,JD,KD),AL(ID,JD,KD),XPL(1500,5)
ENDCLICHE
```

```
      CLICHE NAME18
      USE NAMEB
      DIMENSION QN(ID)
      ENDCLICHE
      CLICHE NAME19
      USE NAMEB
      DIMENSION CNOR(101),DJ(ID,KD),DJM(ID,KD),V3(ID,KD),V4(ID,KD)
     1,V5(ID,KD),V6(ID,KD)
      ENDCLICHE
      CLICHE NAME20
      COMMON /TEST/ EMMODE,ENMODE,EAMC1,EAMC2,EAMC3,EAMC4,EAMC5,EAMC6
      ENDCLICHE
      CLICHE NAME21
      USE NAMEB
      COMMON/PHI/ ABI(11,7,ID),BBI(11,7,ID),CBI(11,7,ID),DBI(11,7,ID),
     1AOL(11,7,ID),BOL(11,7,ID),COL(11,7,ID),DOL(11,7,ID),ABL(11,7,ID),
     1BBL(11,7,ID),CBL(11,7,ID),DBL(11,7,ID),AMERN(ID),ABJN(ID),ABZN(ID)
     1,ARL(11,7,ID),BRL(11,7,ID),CRL(11,7,ID),DRL(11,7,ID)
     1,PHU1(ID,KD),PHV1(ID,KD),PHD(ID,KD),PHB2(ID,KD)
     1,PPRIM(ID),AWN(ID),BS2(ID),BS4(ID)
     1,BSAV(ID),TRAP(ID,11),SNLP(ID,11),BMAGM(11),NMO
      ENDCLICHE
      CLICHE NAME22
      COMMON/AUX/ U1(JD,KD),V1(JD,KD),B2(JD,KD),GL(JD,KD),V3(JD,KD),
     1D(JD,KD),PHI(JD,KD),PSI(JD,KD),PGL(ID,KD),PV3(ID,KD),PHPC(JD),
     1PHTC(KD),BB(11,7,4),OL(11,7,4),BL(11,7,4),CL(4),RL(11,7,4)
     1,FILJ(JD),FILK(KD)
      ENDCLICHE
      CLICHE NAME23
      DIMENSION NRES(50),PMU(200),PLA(200),FP(200),FPP(200),
     1FPPP(200),OPP1(200),OPP2(200),OPP3(200),EL(ID,KD),GL(ID,KD),
     1GM(ID,KD),XUP(ID,KD),XVP(ID,KD),AWW(50),YUP(3,JD),YVP(3,KD),
     1ABBF(101),ABJF(101),AZZF(101)
      ENDCLICHE
C
      PROGRAM BETA (INPUT=65,OUTPUT=514,TAPE3=514,TAPE4=514,TAPE2=65
     1 ,TAPE1=514,TAPE25,TAPE7)
C     THIS MAIN ROUTINE CALLS OTHER SUBROUTINES,CONTROLS INPUT-OUTPUT,
C     AND COMPUTES FOURIER COEFFICIENTS
      COMMON /PRINT/ IX(8), IJ(8), JX(50), PJX(50), NNJ, NJX
      USE NAME1
      USE NAME2
      USE NAME6
      USE NAME7
      USE NAME8
      USE NAME9
      USE NAME10
      USE NAME11
      USE NAME12
      USE NAME13
      USE NAME14
      USE NAME15
      USE NAME16
```

```
      USE NAME18
      USE NAME20
      USE NAME21
      CALL LINK ("UNIT25=(DATF1,OPEN),UNIT6=(RUNSAMP,CREATE,TEXT),
     1 UNIT003=(T123,CREATE,     ),UNIT004=(T124,CREATE,     ),
     1UNIT002=(T122,CREATE,     ),
     2 UNIT001=(T121,CREATE,     ),UNIT007=(T127,CREATE,     ),READ25,
     3 PRINT6//")
      DT1=0.
C     READ INPUT DATA
      READ (25,1720)
      READ (25,1670) EP,RBOU,QLZ,NRUN,NGEOM,NUMB
      RX=1.0
      RY=RBOU
      READ (25,1720)
      IF (NGEOM.GT.3) GO TO 10
      READ (25,1680) DEL0,DEL1,DEL2,DEL3,DEL10,DEL20,DEL30
      READ (25,1720)
      READ (25,1680) DEL22,DEL33,DELA,DELB,DELC
      GO TO 20
   10 CONTINUE
      READ (25,1680) DEL1R,DEL1Z,DEL(1,1),DEL(2,1),DEL(3,1),DEL(1,2),DEL
     1 (2,2)
      READ (25,1720)
      READ (25,1680) DEL(3,2),DEL(1,3),DEL(2,3),DEL(3,3),DEL(1,4),DEL(2,
     1 4),DEL(3,4)
   20 CONTINUE
      READ (25,1720)
      READ (25,1690) XPR,YPR,ZPR,P0,AMU0,AMU1,AMU2
      READ (25,1720)
      READ (25,1700) ALF,ALFU
      READ (25,1720)
      READ (25,1610) NI,NJ,NK,ASYE,ERR
      READ (25,1720)
      READ (25,1600) SA1,SA2,SA3,DT
      READ (25,1720)
      READ (25,1620) SE1,SEMU,SEAX
      READ (25,1720)
      READ (25,1630) SAFI,NE,NVAC
      INVAC=IABS(NVAC)
      READ (25,1720)
      READ (25,1640) NR,NZ,NT,NAC,DPSI,GAM,IGAM
      READ (25,1720)
      READ (25,1510) MIS1,NIS,NAS,IROT,MF,NF,IPF
      MIS=IABS(MIS1)
      READ (25,1720)
      READ (25,1650) IC,ITERF,TLIM
      NP=1
      READ (25,1720)
      READ (25,1720) (IX(I),I=1,8)
      READ (25,1720)
      READ (25,1460) US0,UIN,USM,IVAX
      READ (25,1720)
```

```
      READ (25,1660) NIV,NV,NP,OM,SAFV,SE4,C1,C2
      READ (25,1720)
      READ (25,1470) KW1,KW2,AK3,TORS,AMPH,PO1,WRAD,VERT
      READ (25,1720)
      READ (25,1730) NRA1,NRA2,NZA1,NZA2,MK1,MK2,MK3,MK4
      READ (25,1720)
      READ (25,1621) EL1,EL2,EM1,EM2,EMMODE,ENMODE
      READ (25,1720)
      READ (25,1680) EAMC1,EAMC2,EAMC3,EAMC4,EAMC5,EAMC6
C     PRINT INPUT DATA
      PRINT 1740
      PRINT 1750, EP,RBOU,QLZ,NRUN,NGEOM,NUMB
      IF (NGEOM.EQ.4) GO TO 50
      PRINT 1780, DEL0,DEL1,DEL2,DEL3,DEL10,DEL20,DEL30
      PRINT 1790, DEL22,DEL33,DELA,DELB,DELC
      GO TO 60
   50 CONTINUE
      PRINT 1760, DEL1R,DEL1Z,DEL(1,1),DEL(2,1),DEL(3,1),DEL(1,2),DEL(2,
     1 2)
      PRINT 1770, DEL(3,2),DEL(1,3),DEL(2,3),DEL(3,3),DEL(1,4),DEL(2,4)
     1 ,DEL(3,4)
   60 CONTINUE
      PRINT 1800, XPR,YPR,ZPR,PO,AMU0,AMU1,AMU2
      PRINT 1810, ALF,ALFU
      PRINT 1820, NI,NJ,NK,ASYE,ERR
      PRINT 1830, SA1,SA2,SA3,DT
      PRINT 1840, SE1,SEMU,SEAX
      PRINT 1850, SAFI,NE,NVAC
      PRINT 1710, NR,NZ,NT,NAC,DPSI,GAM,IGAM
      PRINT 1520, MIS1,NIS,NAS,IROT,MF,NF,IPF
      MF=MF+1
      NF=NF+1
      PRINT 1860, IC,ITERF,TLIM
      PRINT 1880, (IX(I),I=1,8)
      PRINT 1500, US0,UIN,USM,IVAX
      PRINT 1870, NIV,NV,NP,OM,SAFV,SE4,C1,C2
      PRINT 1490, KW1,KW2,AK3,TORS,AMPH,PO1,WRAD,VERT
      PRINT 1890, NRA1,NRA2,NZA1,NZA2,MK1,MK2,MK3,MK4
      PRINT 1891, EL1,EL2,EM1,EM2,EMMODE,ENMODE
      PRINT 1892, EAMC1,EAMC2,EAMC3,EAMC4,EAMC5,EAMC6
  101 CONTINUE
C     DEFINE CONSTANTS
      PI=3.1415926535898
      AMAGM=0.66
      FAXIS=1.0
      FAXV=1.0
      ELIM=2.0
      SAFAX=1.000
      IT=0
      I1=(NI-1)/2+1
      IO=50*IC
      N1=NJ+1
      N2=NJ+2
```

```
      N3=NI-1
      N4=NK+1
      N5=NK+2
      NJA=5
      IF (NVAC.LT.0) NJA=3
      IF (EP.LT.0.00001) GO TO 110
      ZLE=(2.0*PI)/(EP*QLZ)
      GO TO 120
  110 ZLE=QLZ
  120 SM=1.0
      RMN=FLOAT(NIS)/FLOAT(MIS)
      HS=SM/N3
      NPLOT=-1
      HU=1.0/NJ
      HV=1.0/NK
      IF (NVAC.LT.0) GO TO 130
      N6=NIV-1
      HR=1.0/N6
      H1=1.0/(HR*HR)
      H2=1.0/(HU*HU)
      H3=1.0/(HR*HU)
      HR4=0.25*H1
      DTB=DT/SE4
      PM=1.0-OM
  130 CONTINUE
      A1=EP/8.0
      A2=1.0/(4.0*HS)
      A3=1.0/(4.0*HU*HU*ZLE*ZLE)
      A4=1.0/(4.0*HV*HV*ZLE*ZLE)
      A5=1.0/(4.0*HU*HV*ZLE*ZLE)
      A6=1.0/(8.0*HS)
      HU4=1.0/(4.0*HU*HU)
      HV4=1.0/(4.0*HV*HV)
      HUV=1.0/(4.0*HU*HV)
      RA1=1.0/(DT*DT)
      RE1=SE1/(SA1*DT)
      RA2=RA1
      RA3=RA1
      RE2=RE1
      RE3=RE1
      E4=EP/4.0
      NE1=NE-1
C     DEFINE MATRICES FOR FOURIER ANALYSIS
      DO 140 K=1,N5
      V=2.0*PI*(K-2.5)*HV
      VP1=2.0*PI*(K-2.0)*HV
      SVP(1,K)=1.0
      SVP(2,K)=SIN(VP1)
      SVP(3,K)=COS(VP1)
      SVP(4,K)=SIN(2.0*VP1)
      SVP(5,K)=COS(2.0*VP1)
      SVP(6,K)=SIN(3.0*VP1)
      SVP(7,K)=COS(3.0*VP1)
```

```
       SV(1,K)=1.0
       SV(2,K)=SIN(V)
       SV(3,K)=COS(V)
       SV(4,K)=SIN(2.0*V)
       SV(5,K)=COS(2.0*V)
       SV(6,K)=SIN(3.0*V)
  140  SV(7,K)=COS(3.0*V)
       DO 150 J=1,N2
       U=2.0*PI*(J-2.5)*HU
       UP1=2.0*PI*(J-2.0)*HU
       SUP(1,J)=1.0
       SUP(2,J)=SIN(UP1*(1.0+UIN))
       SUP(3,J)=COS(UP1*(1.0+UIN))
       SUP(4,J)=SIN((2.0+UIN)*UP1)
       SUP(5,J)=COS((2.0+UIN)*UP1)
       SUP(6,J)=SIN((3.0+UIN)*UP1)
       SUP(7,J)=COS((3.0+UIN)*UP1)
       SU(1,J)=1.0
       SU(2,J)=SIN(U*(1.0+UIN))
       SU(3,J)=COS(U*(1.0+UIN))
       SU(4,J)=SIN((2.0+UIN)*U)
       SU(6,J)=SIN((3.0+UIN)*U)
       SU(7,J)=COS((3.0+UIN)*U)
  150  SU(5,J)=COS((2.0+UIN)*U)
       SEAXI=SEAX
       IF (SEAX.LT.0.0) SEAX=3.0*SEAXI
       ITER=0
       ITER1=ITER
       NAC1=NAC
       NTER=100000
       KTER=100000
       IRQ=-1
       RQ=0.0
       CALL ASYM (ASYE)
       CALL SURF
       CALL PRNT
       PRINT 1480
C      SET INITIAL TIME DERIVATIVE FOR R AND PSI EQUAL TO ZERO
       ANORM=0.0
       BNORM=0.0
       DO 160 I=1,NI
       OP1(I)=0.0
       OP3(I)=0.0
  160  OP2(I)=0.0
       DO 180 K=1,N5
       RPAX(K)=ROAX(K)
       RPBX(K)=ROBX(K)
       RPCX(K)=ROCX(K)
       DO 170 J=1,N2
       DO 170 I=1,NI
       XO(I,J,K)=RO(I,J,K)
  170  XL(I,J,K)=AL(I,J,K)
  180  CONTINUE
```

```
      IF (INVAC.LT.10) GO TO 200
      DO 190 I=1,NI
  190 QN(I)=Q(I)
      RA5=SAMU/(DT*DT)
      RE5=SEMU/DT
      DG5=1.0/(RA5+RE5)
      DH5=2.0*RA5+RE5
  200 CONTINUE
      IF (NT.GT.0) GO TO 210
      GO TO 240
  210 CONTINUE
      RIN=0.0
      ZIN=0.0
C     COMPUTATION OF AXIS FOURIER COEFFICIENTS WHICH WILL BE FIXED
      DO 220 K=2,N4
      RIN=RIN+RA(K)*SV(NR,K)
  220 ZIN=ZIN+ZA(K)*SV(NZ,K)
      RIN=RIN*2.0*HV
      ZIN=ZIN*2.0*HV
      IF (NR.EQ.1) RIN=0.5*RIN
      IF (NZ.EQ.1) ZIN=0.5*ZIN
      IF (NVAC.LT.0) GO TO 240
      RIN1=0.0
      ZIN1=0.0
      DO 230 K=2,N4
      RIN1=RIN1+RVA(K)*SV(NR,K)
  230 ZIN1=ZIN1+ZVA(K)*SV(NZ,K)
      RIN1=RIN1*2.0*HV
      ZIN1=ZIN1*2.0*HV
      IF (NR.EQ.1) RIN1=0.5*RIN1
      IF (NZ.EQ.1) ZIN1=0.5*ZIN1
  240 CONTINUE
C     INITIAL VALUES FOR DESCENT COEFFICIENTS
      DO 250 I=1,NE
      OP(I)=0.0
      OE(I)=SE1/SA1
  250 EN(I)=OE(I)*OE(I)/(SAFI*SAFI)
      IF (NVAC.LT.0) GO TO 260
      OM1=OM
      OM2=OM
      VERT=FAXV
      VERTO=VERT
      PO2=0.0
  260 CONTINUE
      DT2=2.0*DT
      RA1=1.0/(DT*DT)
      RA2=RA1
      RA3=RA1
      ETOT1=0.0
      EVNE=0.0
      ERBO=0.0001
      ERBO1=0.0001
      ERVA=0.0001
```

```
      ERMU=0.0000001
      INI=0
      RE1=SE1/(SA1*DT)
      RE2=RE1
      RE3=RE1
      DO 270 J=1,5
  270 AA2(J)=0.0
      REWIND 3
      REWIND 4
C     SET INITIAL TIME DERIVATIVE FOR MAGNETIC AXIS EQUAL TO ZERO
      DO 280 K=1,N5
      RN(K)=RA(K)
  280 ZN(K)=ZA(K)
  290 CONTINUE
      STER=FLOAT(ITER-ITER1)
      X1=1.0+19.0*(2.0-STER/NAC1)
      LTER=ITER-ITER1
      IF (LTER.LT.NAC1) X1=20.0
      EAC=AMAX1(1.0,X1)
      IF (NVAC.LT.0) GO TO 330
      CALL TSOR (ERVA,EVNE,IT,ITER)
      IF (IRQ.GT.0) GO TO 330
      IF (PO1.LT.0.0) GO TO 330
      DVERT=DT*PO2/(PO1*EAC)
C     VERTICAL FIELD ITERATION
      VERT=VERT+DVERT
      DO 310 J=2,N1
      DO 300 K=2,N4
  300 PT(NIV,J,K)=PT(NIV,J,K)+DVERT*0.25*(Z(J,K)+Z(J-1,K)+Z(J,K-1)+Z(J-1
     1 ,K-1))
      PT(NIV,N2,K)=PT(NIV,2,K)+C1
  310 PT(NIV,1,K)=PT(NIV,N1,K)-C1
      DO 320 J=1,N2
      PT(NIV,J,N5)=PT(NIV,J,2)+C2
  320 PT(NIV,J,1)=PT(NIV,J,N4)-C2
  330 CONTINUE
      NCO=0
  340 CONTINUE
      DG1=1.0/(RA1+RE1)
      DG2=1.0/(RA2+RE2)
      DG3=1.0/(RA3+RE3)
      DH1=2.0*RA1+RE1
      DH2=2.0*RA2+RE2
      DH3=2.0*RA3+RE3
      IF (NVAC.GT.0) DTB=DT/(SE4*EAC)
C     COMPUTE ONE ITERATION OF PLASMA EQUATIONS
      IF (NVAC.LT.10) GO TO 350
  350 CONTINUE
      CALL TGRAD (BJA,SJA,EAX,ERO,EAL,ERBO,EROAX)
      ND=IABS(NVAC)
      IF (ND.LT.10) GO TO 380
      IF (ITER.LT.NAC) GO TO 380
      NTT=-NT
```

```
      IF (NT.LT.0.AND.ITER.GT.NTT) GO TO 380
  360 CONTINUE
      ERMU=0.0
      EDT=DT/SEMU
      DO 370 I=2,NI
      X1=+TC(I)/SL1(I)
      ERMU=AMAX1(ERMU,ABS(X1))
  370 Q(I)=Q(I)+EDT*X1
  380 IF (NT.GT.0) GO TO 820
      NT1=-NT
      FPSI=0.0
      FPSI1=0.0
      IF (ITER.LT.NT1) GO TO 880
      IF (ITER.GT.NT1) GO TO 620
      GAM1=1.0/GAM
      ANORM=0.0
      BNORM=0.0
      PR(1)=PRP(1)
      PR(NI)=PRM(NI)
      OP3(1)=SL2(2)/3.0
      OP3(NI)=1.0-(1.0-SL2(N3))/3.0
      DO 390 I=2,N3
      OP3(I)=0.5*(SA(I)*(SL2(I)+(SL2(I+1)-SL2(I))/3.0)+SA(I-1)*(SL2(I)-
     1 (SL2(I)-SL2(I-1))/3.0))
      OP3(I)=OP3(I)/PQS(I)
  390 PR(I)=0.5*(PRP(I)+PRM(I))
      SRHO=0.0
      DO 400 I=1,NI
      OP1(I)=(PR(I)+0.000001)**GAM1
      IF (IGAM.LT.0) OP1(I)=1.0
      SRHO=SRHO+OP1(I)*PQS(I)
      X1=(2.0*PI*Q(I))/(QT(I)*ZLE)
      X1=X1*RBOU
      OP2(I)=4.0/(1.0+X1*X1*SL2(I))
      X2=1.0-SL1(I)
      BNORM=BNORM+OP1(I)*SL2(I)*PQS(I)
  400 ANORM=ANORM+0.5*OP1(I)*X2*X2*PQS(I)
      SRHO=SRHO*HS
      RBO2=RBOU*RBOU
      RBO4=RBO2*RBO2
      ANORM=ANORM*HS*RBO2/SRHO
      BNORM=BNORM*HS*RBO2/SRHO
      NPLOT=1
      IF (NVAC.GT.0) CALL TSOR (ERVA,EVNE,1,ITER)
      CALL TGRAD (X1,X2,X3,X4,X5,X6,X7)
      NPLOT=-1
      CALL FPRINT (ITER,-1)
      DO 410 I=1,NI
  410 OP1(I)=OP1(I)/SRHO
      PNORM=0.0
      PNORM1=0.0
      PRO=0.0
      IF (NRUN.LE.1) GO TO 580
```

```
          CALL EXTEN
 580  CONTINUE
          DO 590 K=2,N4
          X1=EL1*EH1(K)+EL2*EH2(K)
          X2=EM1*EJ1(K)+EM2*EJ2(K)
 590  PNORM1=PNORM1+X1*X1+X2*X2
          PNORM1=PNORM1*HV*ANORM
          PMA=(ANORM*PNORM1)/SA3
          DO 600 I=1,NI
          X4=OP1(I)*SL2(I)
          X3=PQS(I)*X4
          X5=X3*X4
          X6=SA1*SL2(I)
          X7=OP2(I)*OP2(I)/SA2
          IF (I.EQ.1) X6=SL1(2)
          DO 600 J=2,N1
          DO 600 K=2,N4
          X1=EAM1(I)*EF1(J,K)+EAM2(I)*EF2(J,K)+EAM3(I)*EF3(J,K)
          X2=EAM4(I)*EG1(J,K)+EAM5(I)*EG2(J,K)+EAM6(I)*EG3(J,K)
          X2=-X2
          X1=X1*RBO2
          PNORM=PNORM+X3*(X1*X1+OP2(I)*X2*X2)
 600  PRO=PRO+X5*((X1*X1)/X6+X7*X2*X2)
          PNORM=(PNORM*HS*HU*HV+PNORM1)*PI
          IF (NVAC.LT.0) GO TO 620
          PNORM2=0.0
          DO 610 J=2,N1
          DO 610 K=2,N4
 610  PNORM2=PNORM2+EBD(J,K)*EBD(J,K)
          PNORM2=PNORM2*HU*HV*BNORM
          PNORM=PNORM+PI*PNORM2
          QNORM=PI*BNORM*PNORM2
 620  CONTINUE
          DO 630 K=2,N4
          X1=EL1*EH1(K)+EL2*EH2(K)
          X2=EM1*EJ1(K)+EM2*EJ2(K)
 630  FPSI1=FPSI1+RA(K)*X1+ZA(K)*X2
          FPSI1=FPSI1*HV*ANORM
          DO 640 I=1,NI
          X3=PQS(I)*OP1(I)*SL2(I)
          DO 640 J=2,N1
          U=(J-2.5)*HU
          DO 640 K=2,N4
          V=(K-2.5)*HV
          X1=EAM1(I)*EF1(J,K)+EAM2(I)*EF2(J,K)+EAM3(I)*EF3(J,K)
          X2=EAM4(I)*EG1(J,K)+EAM5(I)*EG2(J,K)+EAM6(I)*EG3(J,K)
          X2=-X2
          X1=X1*RBO4
 640  FPSI=FPSI+X3*(X1*RO(I,J,K)+OP2(I)*X2*(AL(I,J,K)+QT(I)*U-Q(I)*V))
          FPSI=(FPSI*HS*HU*HV+FPSI1)*PI
          IF (NVAC.LT.0) GO TO 660
          FPSI2=0.0
          DO 650 J=2,N1
```

```
          DO 650 K=2,N4
  650 FPSI2=FPSI2+EBD(J,K)*X(J,K)
          FPSI=FPSI+FPSI2*HU*HV*BNORM*PI
  660 CONTINUE
          IF (ITER.NE.NT1) GO TO 810
          ENERZ=ETOT
          DPSI1=DPSI+FPSI
          FPSIO=FPSI
          REWIND 1
          WRITE (1) (((RO(I,J,K),I=1,NI),J=1,N2),K=1,N5),(((AL(I,J,K),I=1,NI
        1 ),J=1,N2),K=1,N5),(RA(K),K=1,N5),(ZA(K),K=1,N5)
          IF (NVAC.GT.0) WRITE (1) ((X(J,K),J=1,N2),K=1,N5),(RVA(K),K=1,N5),
        1 (ZVA(K),K=1,N5)
          X1=DPSI/PNORM
          CALL PROJ(X1,ITER)
          ITER1=ITER
          RE1=SE1/(SA1*DT)
          RE2=RE1
          RE3=RE1
          PNORM=(PRO*HS*HU*HV+PMA)*PI
          DO 800 I=1,NE
          OP(I)=0.0
          OE(I)=SE1/SA1
  800 EN(I)=OE(I)*OE(I)/(SAFI*SAFI)
          FPSI=DPSI1
          SAFAX=1.000
          NTER=100000
          KTER=100000
          IF (SEAX.LT.0.0) SEAX=3.0*SEAXI
          IRQ=1
  810 IF (ITER.LT.NT1) GO TO 880
          FPSI=DPSI1-FPSI
          GO TO 880
  820 CONTINUE
          NT1=NT+10
          IF (ITER.GT.NT1) GO TO 850
C       FIXES GIVEN AXIS FOURIER COEFFICIENTS
          ALF=(1.0+NT1-ITER)/10.0
          IF (ALF.GT.1.0) ALF=1.0
          SUM1=0.0
          SUM2=0.0
          DO 830 K=2,N4
          SUM1=SUM1+RA(K)*SV(NR,K)
  830 SUM2=SUM2+ZA(K)*SV(NZ,K)
          SUM1=SUM1*2.0*HV
          SUM2=SUM2*2.0*HV
          IF (NR.EQ.1) SUM1=0.5*SUM1
          IF (NZ.EQ.1) SUM2=0.5*SUM2
          DO 840 K=1,N5
          RA(K)=RA(K)+ALF*SV(NR,K)*(RIN-SUM1)
  840 ZA(K)=ZA(K)+ALF*SV(NZ,K)*(ZIN-SUM2)
  850 CONTINUE
          IF (NVAC.LT.0) GO TO 880
```

```
          IF (ITER.GT.NT1) GO TO 880
          SUM1=0.0
          SUM2=0.0
          DO 860 K=2,N4
          SUM1=SUM1+RVA(K)*SV(NR,K)
      860 SUM2=SUM2+ZVA(K)*SV(NZ,K)
          SUM1=SUM1*2.0*HV
          SUM2=SUM2*2.0*HV
          IF (NR.EQ.1) SUM1=0.5*SUM1
          IF (NZ.EQ.1) SUM2=0.5*SUM2
          DO 870 K=1,N5
          RVA(K)=RVA(K)+ALF*SV(NR,K)*(RIN1-SUM1)
      870 ZVA(K)=ZVA(K)+ALF*SV(NZ,K)*(ZIN1-SUM2)
      880 CONTINUE
          RATIO=SJA/BJA
          IF (RATIO.GT.200.0) GO TO 1220
          NCO=NCO+1
          IF (NCO.LT.NP) GO TO 340
          IF (NVAC.LT.0) GO TO 890
          IF (ITER.LT.50) GO TO 890
          CALL TBO (ERBO,ERBO1,IRQ)
      890 CONTINUE
          ITER=ITER+1
          IF (ITER.EQ.6*NT) SEAX=SEAX/1.
          ETOT=ENER+EVNE
          GEN=ETOT-ETOT1
          ETOT1=ETOT
          EVAC=EVNE
          IF (DT1.EQ.0.) GO TO 900
          IF (DT1.EQ.DT) GO TO 900
          DT=DT1
          DT2=2.*DT
          IF (NVAC.GT.0) DTB=DT/SE4
          RA1=1.0/(DT*DT)
          RA2=RA1
          RA3=RA1
      900 CONTINUE
          NTER=NTER+1
          KTER=KTER+1
C         NEW VALUES FOR DESCENT COEFFICIENTS
          DO 910 I=1,NE1
      910 OP(I)=OP(I+1)
          OP(NE)=AA1(1)+AA1(2)+AA1(3)
          IF (ITER.LT.NAC) GO TO 930
          DO 920 I=1,NE1
      920 EN(I)=EN(I+1)
          CALL ELAM
      930 CONTINUE
          DO 940 J=1,NJA
          AA3(J)=AA2(J)
      940 AA2(J)=AA1(J)
          IF (ITER.LT.NAC) GO TO 970
          SUM=0.0
```

```
         DO 950 I=1,NE
  950 SUM=SUM+EN(I)
      SUM=SQRT(SUM/NE)
      DO 960 I=1,NE1
  960 OE(I)=OE(I+1)
      OE(NE)=SUM*SAFI
      RE1=OE(NE)/DT
      RE2=RE1
      RE3=RE1
      SAFAX=1.0
      IF (SEAX.LT.0.0) SEAX=SEAXI
      IF (NVAC.LT.0) GO TO 970
      OM1=OM
      OM2=OM
  970 CONTINUE
C     PRINTOUT, TIME AND ERROR CRITERIA
      IF (KTER.GE.IO) GO TO 1000
  980 IF (NTER.GE.IC) GO TO 1010
  990 TMAX=AMAX1(EAX,ERO,EAL,ERBO)
      CALL SECOND (TIM1)
      IF (ITER.GT.ITERF) GO TO 1220
      IF (TIM1.GE.TLIM) GO TO 1220
      IF (TMAX.GT.ERR) GO TO 290
      GO TO 1220
 1000 KTER=0
      INI=0
      GO TO 980
 1010 NTER=0
C     HEADING FOR ITERATION DATA PRINTED OUT
      IF (INI.EQ.0) PRINT 1900, (JX(IJ(J)),J=1,NNJ)
      IF (INI.EQ.0) PRINT 1910
      INI=1
      CALL SECOND (TIM1)
C     COMPUTATION OF FOURIER COEFFICIENTS
      DO 1030 I=1,7
      XR(I)=0.0
      XZ(I)=0.0
      DO 1020 K=2,N4
      XR(I)=XR(I)+RA(K)*SV(I,K)
 1020 XZ(I)=XZ(I)+ZA(K)*SV(I,K)
      XR(I)=2.0*HV*XR(I)
 1030 XZ(I)=2.0*HV*XZ(I)
      XR(1)=0.5*XR(1)
      XZ(1)=0.5*XZ(1)
      WRITE (3) ETOT,ERO,EAL,EAX
      WRITE (3) (XR(I),I=1,7),(XZ(I),I=1,7)
      IF (NVAC.LT.0) GO TO 1060
      DO 1050 I=1,7
      YR(I)=0.0
      YZ(I)=0.0
      DO 1040 K=2,N4
      YR(I)=YR(I)+RVA(K)*SV(I,K)
 1040 YZ(I)=YZ(I)+ZVA(K)*SV(I,K)
```

```
        YR(I)=2.0*HV*YR(I)
 1050   YZ(I)=2.0*HV*YZ(I)
        YR(1)=0.5*YR(1)
        YZ(1)=0.5*YZ(1)
        WRITE (4) ERBO,ERBO1,ERVA
        WRITE (4) (YR(I),I=1,7),(YZ(I),I=1,7)
 1060   CONTINUE
        IF (NVAC.GT.0) GO TO 1130
        DO 1090 K=2,N4
        DO 1080 L=1,7
        SFI(L,K)=0.0
        DO 1070 J=2,N1
 1070   SFI(L,K)=SFI(L,K)+RO(I1,J,K)*SU(L,J)
 1080   SFI(L,K)=SFI(L,K)*2.0*HU
 1090   SFI(1,K)=0.5*SFI(1,K)
        DO 1120 L=1,7
        DO 1110 M=1,7
        SRO(L,M)=0.0
        DO 1100 K=2,N4
 1100   SRO(L,M)=SRO(L,M)+SFI(L,K)*SV(M,K)
 1110   SRO(L,M)=SRO(L,M)*2.0*HV
 1120   SRO(L,1)=0.5*SRO(L,1)
        GO TO 1200
 1130   CONTINUE
        DO 1160 K=2,N4
        V=(K-2)*HV
        DO 1150 L=1,7
        SFI(L,K)=0.0
        DO 1140 J=2,N1
 1140   SFI(L,K)=SFI(L,K)+X(J,K)*SU(L,J)
 1150   SFI(L,K)=SFI(L,K)*2.0*HU
 1160   SFI(1,K)=0.5*SFI(1,K)
        DO 1190 L=1,7
        DO 1180 M=1,7
        SRO(L,M)=0.0
        DO 1170 K=2,N4
 1170   SRO(L,M)=SRO(L,M)+SFI(L,K)*SV(M,K)
 1180   SRO(L,M)=SRO(L,M)*2.0*HV
 1190   SRO(L,1)=0.5*SRO(L,1)
 1200   CONTINUE
        Y00=SRO(1,1)
        Y10=SQRT(SRO(2,1)*SRO(2,1)+SRO(3,1)*SRO(3,1))
        Y20=SQRT(SRO(4,1)*SRO(4,1)+SRO(5,1)*SRO(5,1))
        Y30=SQRT(SRO(6,1)*SRO(6,1)+SRO(7,1)*SRO(7,1))
        Y01=SQRT(SRO(1,2)*SRO(1,2)+SRO(1,3)*SRO(1,3))
        Y11=SQRT(SRO(2,2)*SRO(2,2)+SRO(2,3)*SRO(2,3)+SRO(3,2)*SRO(3,2)+SRO
       1 (3,3)*SRO(3,3))
        Y21=SQRT(SRO(4,2)*SRO(4,2)+SRO(4,3)*SRO(4,3)+SRO(5,2)*SRO(5,2)+SRO
       1 (5,3)*SRO(5,3))
        Y31=SQRT(SRO(6,2)*SRO(6,2)+SRO(6,3)*SRO(6,3)+SRO(7,2)*SRO(7,2)+SRO
       1 (7,3)*SRO(7,3))
        Y02=SQRT(SRO(1,4)*SRO(1,4)+SRO(1,5)*SRO(1,5))
        Y12=SQRT(SRO(2,4)*SRO(2,4)+SRO(3,4)*SRO(3,4)+SRO(2,5)*SRO(2,5)+SRO
```

```
      1 (3,5)*SRO(3,5))
        Y22=SQRT(SRO(4,4)*SRO(4,4)+SRO(4,5)*SRO(4,5)+SRO(5,4)*SRO(5,4)+SRO
      1 (5,5)*SRO(5,5))
        Y32=SQRT(SRO(6,4)*SRO(6,4)+SRO(7,4)*SRO(7,4)+SRO(6,5)*SRO(6,5)+SRO
      1 (7,5)*SRO(7,5))
        Y03=SQRT(SRO(1,6)*SRO(1,6)+SRO(1,7)*SRO(1,7))
        Y13=SQRT(SRO(2,6)*SRO(2,6)+SRO(3,6)*SRO(3,6)+SRO(2,7)*SRO(2,7)+SRO
      1 (3,7)*SRO(3,7))
        Y23=SQRT(SRO(4,6)*SRO(4,6)+SRO(5,6)*SRO(5,6)+SRO(4,7)*SRO(4,7)+SRO
      1 (5,7)*SRO(5,7))
        Y33=SQRT(SRO(6,6)*SRO(6,6)+SRO(7,6)*SRO(7,6)+SRO(6,7)*SRO(6,7)+SRO
      1 (7,7)*SRO(7,7))
        WRITE (3) Y00,Y10,Y20,Y30,Y01,Y11,Y21,Y31,Y02,Y12,Y22,Y32,Y03,Y13
      1 ,Y23,Y33
        Y1=RE1*DT
        Y2=Y1
        Y3=Y1
        WRITE (3) Y1,EROAX,ERMU
        WRITE (4) OM1,OM2
C       SELECTION OF DATA TO BE PRINTED
        PJX(1)=EAX
        PJX(2)=ERO
        PJX(3)=EAL
        PJX(4)=ERBO
        PJX(5)=ERVA
        PJX(6)=GEN
        PJX(7)=RATIO
        PJX(8)=EROAX
        DO 1210 I=1,7
        PJX(I+8)=XR(I)
 1210   PJX(I+15)=XZ(I)
        PJX(23)=Y00
        PJX(24)=Y10
        PJX(25)=Y20
        PJX(26)=Y30
        PJX(27)=Y01
        PJX(28)=Y11
        PJX(29)=Y21
        PJX(30)=Y31
        PJX(31)=Y02
        PJX(32)=Y12
        PJX(33)=Y22
        PJX(34)=Y32
        PJX(35)=Y03
        PJX(36)=Y13
        PJX(37)=Y23
        PJX(38)=Y33
        PJX(39)=Y1
        PJX(40)=Y2
        PJX(41)=Y3
        PJX(42)=OM1
        PJX(43)=OM2
        PJX(44)=ERMU
```

```
          PJX(45)=A11
          PJX(46)=A22
          PJX(47)=A12
          PJX(48)=VERT
          PJX(49)=ETOT
          PJX(50)=FPSI
          PRINT 1920, ITER,(PJX(IJ(J)),J=1,NNJ)
          GO TO 990
     1220 IF (RATIO.GE.30) PRINT 1930, RATIO
          PRINT 1540, TIM1
          NPLOT=1
C         COMPUTE FINAL VALUES OF MAGNETIC FIELD
          IF (NT.GT.0) GO TO 1230
          IF (ITER.LT.NT1) GO TO 1230
          DELTE=ETOT-ENERZ
     1230 CONTINUE
          CALL TGRAD (BJA,SJA,EAX,ERO,EAL,ERBO,EROAX)
          IF (NVAC.LT.0) GO TO 1240
          CALL TSOR (ERVA,EVNE,IT,ITER)
     1240 CONTINUE
C         FINAL OUTPUT AND PLOTTING
          IF (NT.LT.0) GO TO 1260
          DO 1250 I=1,NI
          DO 1250 J=1,N2
          U=(J-2.5)*HU
          DO 1250 K=1,N5
          V=(K-2.5)*HV
          XO(I,J,K)=RO(I,J,K)
     1250 XL(I,J,K)=AL(I,J,K)+QT(I)*U-Q(I)*V
          GO TO 1320
     1260 CONTINUE
          REWIND 1
C         READ EQUILIBRIUM SOLUTION
          READ (1) (((XO(I,J,K),I=1,NI),J=1,N2),K=1,N5),(((XL(I,J,K),I=1,NI)
         1 ,J=1,N2),K=1,N5),(RN(K),K=1,N5),(ZN(K),K=1,N5)
C         COMPUTE DIFFERENCE BETWEEN STABILITY AND EQUILIBRIUM SOLUTION
          DO 1270 K=1,N5
          RN(K)=RA(K)-RN(K)
          ZN(K)=ZA(K)-ZN(K)
          DO 1270 J=1,N2
          DO 1270 I=1,NI
          XL(I,J,K)=AL(I,J,K)-XL(I,J,K)
     1270 XO(I,J,K)=RO(I,J,K)-XO(I,J,K)
C         COMPUTE EIGENFUNCTION NORM
          SNORM1=0.0
          IF (NVAC.LT.0) GO TO 1290
          READ (1) ((AKB(J,K),J=1,N2),K=1,N5)
          SNORM2=0.0
          DO 1280 J=2,N1
          DO 1280 K=2,N4
          X1=X(J,K)-AKB(J,K)
     1280 SNORM2=SNORM2+X1*X1
          SNORM2=PI*ZLE*SNORM2*BNORM*HU*HV
```

```
 1290 CONTINUE
      DO 1300 K=2,N4
 1300 SNORM1=SNORM1+RN(K)*RN(K)+ZN(K)*ZN(K)
      SNORM1=SNORM1*HV
      SNORM=0.0
      DO 1310 I=1,NI
      X1=PQS(I)*OP1(I)*SL2(I)
      DO 1310 J=2,N1
      DO 1310 K=2,N4
 1310 SNORM=SNORM+X1*(XO(I,J,K)*XO(I,J,K)*RBO4+OP2(I)*XL(I,J,K)*XL(I,J,K
    1 ))
      SNORM=(SNORM*HS*HU*HV+SNORM1*ANORM)*PI*ZLE
      IF (NVAC.GT.0) SNORM=SNORM+SNORM2
      SNORM=0.5*SNORM
      EVAL=DELTE/SNORM
      IF (NVAC.GT.0) EVAL=EVAL*RBOU*RBOU
 1320 CONTINUE
C     FOURIER ANALYSIS OF THE SOLUTION
      DO 1440 I=1,NI
      IXL=1
 1330 CONTINUE
      DO 1360 K=2,N4
      DO 1350 L=1,7
      SFI(L,K)=0.0
      DO 1340 J=2,N1
 1340 SFI(L,K)=SFI(L,K)+XL(I,J,K)*SU(L,J)
 1350 SFI(L,K)=SFI(L,K)*2.0*HU
 1360 SFI(1,K)=0.5*SFI(1,K)
      DO 1390 L=1,7
      DO 1380 M=1,7
      STO(L,M)=0.0
      DO 1370 K=2,N4
 1370 STO(L,M)=STO(L,M)+SFI(L,K)*SV(M,K)
      STO(L,M)=STO(L,M)*2.0*HV
 1380 CONTINUE
 1390 STO(L,1)=0.5*STO(L,1)
      IF (IXL.LT.0) GO TO 1420
      DO 1400 K=2,N4
      DO 1400 J=2,N1
 1400 XL(I,J,K)=XO(I,J,K)
      DO 1410 L=1,7
      DO 1410 M=1,7
 1410 XO(I,L,M)=STO(L,M)
      IXL=-1
      GO TO 1330
 1420 CONTINUE
      DO 1430 L=1,7
      DO 1430 M=1,7
 1430 XL(I,L,M)=STO(L,M)
 1440 CONTINUE
C     STORE RESULTS
      REWIND 7
      WRITE (7) NVAC,NI,NJ,NK,NUMB
```

```
      WRITE (7) HS,HU,HV,EP,QLZ
      WRITE(7) ((R(J,K),J=1,N2),K=1,N5),((Z(J,K),J=1,N2),K=1,N5)
      WRITE(7) ((RU(J,K),J=1,N2),K=1,N5),((ZU(J,K),J=1,N2),K=1,N5)
      WRITE (7) (RA(K),K=1,N5),(ZA(K),K=1,N5)
      WRITE (7) (((RO(I,J,K),K=1,N5),J=1,N2),I=1,NI)
      WRITE (7) (((AL(I,J,K),K=1,N5),J=1,N2),I=1,NI)
      WRITE (7) (SL1(I),I=1,NI),(QT(I),I=1,NI),(Q(I),I=1,NI)
      IF (NVAC.LT.0) GO TO 1450
      WRITE (7) NIV,HR,C1,C2
      WRITE (7) ((X(J,K),K=1,N5),J=1,N2)
      WRITE (7) (((PT(I,J,K),K=1,N5),J=1,N2),I=1,NIV)
 1450 CONTINUE
      WRITE(7) (SNL2(I),I=1,NI),MF,NF
      WRITE(7) (PR(I),I=2,N3),(PPRIM(I),I=2,N3)
      WRITE(7) (((ABI(M,N,I),M=1,MF),N=1,NF),I=2,N3)
      WRITE(7) (((BBI(M,N,I),M=1,MF),N=1,NF),I=2,N3)
      WRITE(7) (((CBI(M,N,I),M=1,MF),N=1,NF),I=2,N3)
      WRITE(7) (((DBI(M,N,I),M=1,MF),N=1,NF),I=2,N3)
      WRITE(7) (((ABL(M,N,I),M=1,MF),N=1,NF),I=2,N3)
      WRITE(7) (((BBL(M,N,I),M=1,MF),N=1,NF),I=2,N3)
      WRITE(7) (((CBL(M,N,I),M=1,MF),N=1,NF),I=2,N3)
      WRITE(7) (((DBL(M,N,I),M=1,MF),N=1,NF),I=2,N3)
      CALL FPRINT (ITER,1)
      CALL TPLOT (ITER,NISL)
      STOP
C
 1460 FORMAT (3F8.4,I8)
 1470 FORMAT (2I8,6F8.4)
 1480 FORMAT (1H1////6X,10H3D RESULTS//)
 1490 FORMAT (13X,3HKW1,5X,3HKW2,5X,3HAK3,4X,4HTORS,4X,4HAMPH,3X,5HEVERT
     1,4X,4HWRAD,4X,4HVERT,/8X,2I8,6F8.4//)
 1500 FORMAT (13X,3HUSO,5X,3HUIN,5X,3HUSM,4X,4HIVAX/8X,3F8.4,3X,I5//)
 1510 FORMAT (7I8)
 1520 FORMAT (/,13X,3HMIS,5X,3HNIS,5X,3HNAS,4X,4HIROT,6X,"MF",6X,"NF",
     1 5X,"IPF",/11X,7(I5,3X),/)
 1530 FORMAT (///,6X,"WALL PERTURBATION DZ=DW*SIN(V)",3X,"DW=",F6.3,///)
 1540 FORMAT (//,6X,"EXECUTION TIME=",F9.2,//)
 1550 FORMAT (5I5)
 1560 FORMAT (5F12.8)
 1570 FORMAT (10F8.4)
 1580 FORMAT (4E16.8)
 1590 FORMAT (I5,F12.8,2E16.8)
 1600 FORMAT (4F8.4)
 1610 FORMAT (3I8,2E8.2)
 1620 FORMAT (4F8.4)
 1621 FORMAT(4F8.4,2I8)
 1630 FORMAT (F8.4,2I8)
 1640 FORMAT (4I8,2F8.4,I8)
 1650 FORMAT (2I8,F8.1)
 1660 FORMAT (3I8,5F8.4)
 1670 FORMAT (3F8.4,3I8)
 1680 FORMAT (7F8.4)
 1690 FORMAT (8F8.4)
```

```
1700 FORMAT (2F8.4)
1710 FORMAT (14X,2HNR,6X,2HNZ,6X,2HNT,5X,3HNAC,4X,4HDPSI,5X,3HGAM,4X,4H
    1IGAM/8X,2(3X,I5),1X,I7,3X,I5,1X,F7.3,1X,F7.3,3X,I5,//)
1720 FORMAT (10A8)
1730 FORMAT (8I8)
1740 FORMAT (1H1,5(/),6X,39HINPUT DATA AS READ FROM INPUT DATA DECK//)
1750 FORMAT (14X,2HEP,4X,4HRBOU,5X,3HQLZ,4X,4HNRUN,3X,5HNGEOM,5X,
    1 3HRUN,/9X,3(F7.4,1X),2X,I5,3X,I5,3X,I5//)
1760 FORMAT (11X,5HDEL1R3X,5HDEL1Z3X,5HDEL203X,5HDEL303X,5HDEL40,3X,5HD
    1EL21,3X,5HDEL31/9X,7(F7.4,1X)//)
1770 FORMAT (11X,5HDEL41,3X,5HDEL22,3X,5HDEL32,3X,5HDEL42,3X,5HDEL23,3X
    1 ,5HDEL33,3X,5HDEL43/9X,7(F7.4,1X)//)
1780 FORMAT (11X,5H DELO3X,5H DEL13X,5H DEL2,3X,5H DEL33X,5HDEL103X,5HD
    1EL20,3X,5HDEL30/9X,7(F7.4,1X)//)
1790 FORMAT (11X,5HDEL22,3X,5HDEL33,4X,4HDELA,4X,4HDELB,4X,4HDELC/9X,5(
    1F7.4,1X)//)
1800 FORMAT (13X,3HXPR,5X,3HYPR,5X,3HZPR,6X,2HPO,4X,4HAMUO,4X,4HAMU1,4X
    1,4HAMU2,/9X,7(F7.4, 1X)//)
1810 FORMAT (13X,3HALF,4X,4HALFU,/9X,2(F7.4,1X)//)
1820 FORMAT (14X,2HNI,6X,2HNJ,6X,2HNK,4X,4HASYE,5X,3HERR/8X,3(3X,I5),2
    1 (E8.1)//)
1830 FORMAT (13X,3HSA1,5X,3HSA2,5X,3HSA3,6X,2HDT/9X,3(F7.4,1X),F7.4//)
1840 FORMAT (13X,3HSE1,4X,4HSEMU,4X,4HSEAX/9X,3(F7.2,1X)//)
1850 FORMAT (12X,4HSAFI,6X,2HNE,4X,4HNVAC/9X,F7.4,2(3X,I5)//)
1860 FORMAT (14X,2HIC,3X,5HITERF,4X,4HTLIM/11X,I5,1X,I7,F8.1//)
1870 FORMAT (13X,3HNIV,6X,2HNV,6X,2HNP,6X,2HOM,4X,4HSAFV,5X,3HSE4,5X,3H
    1 C1,5X,3H C2/8X,3(3X,I5),5(1X,F7.4)//)
1880 FORMAT (10X,6HPRINT1,2X,6HPRINT2,2X,6HPRINT3,2X,6HPRINT4,2X,6HPRIN
    1T5,2X,6HPRINT6,2X,6HPRINT7,2X,6HPRINT8/8X,8(A8)//)
1890 FORMAT (12X,4HNRA1,4X,4HNRA2,4X,4HNZA1,4X,4HNZA2,5X,3HMK1,5X,3HMK2
    1 ,5X,3HMK3,5X,3HMK4/8X,8(3X,I5)//)
1891 FORMAT(/13X,3HEL1,5X,3HEL2,5X,3HEM1,5X,3HEM2,7X,1HM,7X,1HN/9X,
    14(F7.4,1X),5X,I2,6X,I2//)
1892 FORMAT(/12X,4HEAM1,4X,4HEAM2,4X,4HEAM3,4X,4HEAM4,4X,4HEAM5,4X,4HEA
    1M6/9X,6(F7.4,1X)//)
1900 FORMAT (1H1///6X,4HITER,3X,8(A10,3X))
1910 FORMAT (1H0)
1920 FORMAT (5X,I5,7E13.4,E17.8)
1930 FORMAT (//6X,19HRATIO EXCEEDS LIMIT,2X,F6.2)
     END




     SUBROUTINE ASYM (ASYE)
C    MAIN AXIALLY SYMMETRIC PROGRAM
     USE NAME3
     USE NAME10
     USE NAME11
```

```
        USE NAME12
        USE NAME13
        USE NAME14
        DO 10 J=1,4
     10 AVE(J)=0.0
        NJA=4
        IF (NVAC.LT.0) NJA=3
C       INITIAL VALUES FOR DESCENT COEFFICIENTS
        DO 20 I=1,NE
        OP(I)=0.0
        OE(I)=SE1/SA1
        EN(I)=OE(I)*OE(I)/(SAFI*SAFI)
     20 CONTINUE
        IF (NVAC.LT.0) GO TO 30
        OM1=OM
        PM1=1.0-OM
     30 CONTINUE
        DT2=2.0*DT
        DO 40 J=1,4
     40 AA2(J)=0.0
        INI=0
        X1=SE1/SA1
        RE1=X1/DT
        RE2=RE1
        RE3=RE1
        ELIM=2.0
        ERBO=0.0
        ERBO1=0.0
        EVNE=0.0
        ERVA=0.0
        NE1=NE-1
        HU2=2.0*HU
        IO1=2*IO
        INI=0
        IT=0
        ITER=0
        IC1=2*IC
        NTER=100000
        KTER=100000
        ENER=0.0
        EVAC=0.0
        ETOT=0.0
        CALL ASIN
        DO 50 I=1,NI
        DO 50 J=1,N2
        R1(I,J)=R(I,J)
     50 Z1(I,J)=AL(I,J)
        RAX=RA
        ZAX=ZA
        ROA=1.0/(RBOU*RBOU)
        ROB=0.0
        ROC=1.0/(RBOU*RBOU)
        ROAP=ROA
```

```
      ROBP=ROB
      ROCP=ROC
      OM3=0.1
      PRINT 350
   60 CONTINUE
      DG1=1.0/(RA1+RE1)
      DG2=1.0/(RA2+RE2)
      DG3=1.0/(RA3+RE3)
      DH1=2.0*RA1+RE1
      DH2=2.0*RA2+RE2
      DH3=2.0*RA3+RE3
C     COMPUTE BOUNDARY QUANTITIES
      DO 70 J=2,N1
      X1=0.5*(X(J)+X(J+1))
      X2=(X(J+1)-X(J))/HU
      RB(J)=R2+X1*(RR(J)-R2)
      ZB(J)=Z2+X1*(ZZ(J)-Z2)
      RBU(J)=X2*(RR(J)-R2)+X1*RU(J)
   70 ZBU(J)=X2*(ZZ(J)-Z2)+X1*ZU(J)
      RB(1)=RB(N1)
      ZB(1)=ZB(N1)
      RBU(1)=RBU(N1)
      ZBU(1)=ZBU(N1)
      IF (NVAC.LT.0) GO TO 80
      NCO=0
      CALL ASOR (NCO,ERVA,EVNE,IT)
   80 NCO=0
   90 CONTINUE
      CALL GRAD (ENEW,RMAX,ALMA,AXIE,SJA,BJA)
C     MAGNETIC AXIS ITERATION
      PAS1=DG3*(RA*DH3-GAXR-RA3*RAX)
      PAS2=DG3*(ZA*DH3-GAXZ-RA3*ZAX)
      RAX=RA
      ZAX=ZA
      RA=PAS1
      ZA=PAS2
      X1=(RA-RAX)/DT
      X2=(ZA-ZAX)/DT
      AA1(3)=X1*X1+X2*X2
      AA1(1)=0.0
      AA1(2)=0.0
C     R EQUATION ITERATION
      PAS1=DG1*(ROA*DH1+SUMA-RA1*ROAP)
      PAS2=DG1*(ROB*DH1+SUMB-RA1*ROBP)
      PAS3=DG1*(ROC*DH1+SUMC-RA1*ROCP)
      ROAP=ROA
      ROBP=ROB
      ROCP=ROC
      X1=(ROA-ROAP)/DT
      X2=(ROB-ROBP)/DT
      X3=(ROC-ROCP)/DT
      AA1(3)=AA1(3)+HS*(X1*X1+X2*X2+X3*X3)
      ROA=PAS1
```

```
        ROB=PAS2
        ROC=PAS3
        DO 100 J=2,N1
        X1=0.5*(AF(J)+AF(J-1))
        X2=0.5*(AG(J)+AG(J-1))
        X3=0.5*(AH(J)+AH(J-1))
        X4=ROA*X1+ROB*X2+ROC*X3
        R(1,J)=OM3/SQRT(X4)+(1.0-OM3)*R(1,J)
        DO 100 I=2,N3
        PAS=DG1*(R(I,J)*DH1-GR(I,J)-RA1*R1(I,J))
        X1=(PAS-R(I,J))/DT
        AA1(1)=AA1(1)+X1*X1
        R1(I,J)=R(I,J)
  100   R(I,J)=PAS
C       PSI EQUATION ITERATION
        AL(1,2)=0.0
        SUM1=0.0
        DO 110 J=2,N1
  110   SUM1=SUM1+1.0/(F(1,J)*G(1,J))
        SUM1=-QT(1)/SUM1
        DO 120 J=2,N1
  120   AL(1,J+1)=AL(1,J)+SUM1/(F(1,J)*G(1,J))
        DO 130 J=2,N1
        DO 130 I=2,NI
        PAS=DG2*(AL(I,J)*DH2-GZ(I,J)-RA2*Z1(I,J))
        X1=(PAS-AL(I,J))/DT
        AA1(2)=AA1(2)+X1*X1
        Z1(I,J)=AL(I,J)
  130   AL(I,J)=PAS
        DO 140 I=1,N3
        R1(I,1)=R1(I,N1)
        R(I,1)=R(I,N1)
  140   R(I,N2)=R(I,2)
        DO 150 I=1,NI
        Z1(I,1)=Z1(I,N1)+QT(I)
        AL(I,1)=AL(I,N1)+QT(I)
  150   AL(I,N2)=AL(I,2)-QT(I)
        AA1(1)=AA1(1)*HS*HU
        AA1(2)=AA1(2)*HS*HU
        NCO=NCO+1
        IF (NCO.LT.NP) GO TO 90
        IF (NVAC.LT.0) GO TO 160
        CALL ASBO (ERBO,ERBO1)
  160   CONTINUE
        ITER=ITER+1
        IF (ITER.GE.9000) STOP
        NTER=NTER+1
        DO 170 I=1,NE1
  170   OP(I)=OP(I+1)
        OP(NE)=AA1(1)+AA1(2)+AA1(3)
        IF (ITER.LT.NAC) GO TO 190
C       COMPUTE NEW DESCENT COEFFICIENTS
        DO 180 I=1,NE1
```

```
180 EN(I)=EN(I+1)
    CALL ELAM
190 CONTINUE
    DO 200 J=1,4
    AA3(J)=AA2(J)
200 AA2(J)=AA1(J)
    IF (ITER.LT.NAC) GO TO 230
    SUM=0.0
    DO 210 I=1,NE
210 SUM=SUM+EN(I)
    SUM=SQRT(SUM/NE)
    DO 220 I=1,NE1
220 OE(I)=OE(I+1)
    OE(NE)=SUM*SAFI
    RE1=OE(NE)/DT
    RE2=RE1
    RE3=RE1
    OM3=1.0
230 CONTINUE
    KTER=KTER+1
    PMAX=AMAX1(RMAX,ALMA,AXIE,ERBO)
    SUMABC=AMAX1(ABS(SUMA),ABS(SUMB),ABS(SUMC))
    PMAX=AMAX1(PMAX,SUMABC)
C   PRINTOUT AND ERROR CRITERIA
    IF (KTER.GE.IO1) GO TO 260
240 IF (NTER.GE.IC1) GO TO 270
250 IF (PMAX.LT.ASYE) GO TO 300
    GO TO 60
260 KTER=0
    INI=0
    GO TO 240
270 NTER=0
    GENP=ENEW-ENER
    GENV=EVNE-EVAC
    ENER=ENEW
    EVAC=EVNE
    GEN=GENP-GENV
    ETOT=ENER-EVAC
    IF (INI.NE.0) GO TO 280
    IF (NVAC.LT.0) PRINT 320
    IF (NVAC.GT.0) PRINT 310
280 CONTINUE
    INI=1
C   PRINT ITERATION DATA
    RAT=SJA/BJA
    IF (NVAC.GT.0) GO TO 290
    PRINT 330, ITER,AXIE,RMAX,ALMA,GEN,RAT,ENER
    GO TO 250
290 PRINT 340, ITER,AXIE,RMAX,ALMA,ERVA,ERBO,GEN,RAT,ENER
    GO TO 250
300 CALL ASOUT
    RETURN
C
```

```
  310 FORMAT (1H1///,6X,4HITER,5X,8HAXIS ERR,7X,6HRO ERR,6X,7HPSI ERR,
     16X,7HVAC ERR,6X,7HBOU ERR,6X,7HDELENER,6X,7HJAC RAT,6X,4HENER,/)
  320 FORMAT (1H1///,6X,4HITER,5X,8HAXIS ERR,7X,6HRO ERR,6X,7HPSI ERR,
     16X,7HDELENER,6X,7HJAC RAT,6X,4HENER,/)
  330 FORMAT (5X,I5,5E13.4,E17.10)
  340 FORMAT (5X,I5,7E13.4,E17.10)
  350 FORMAT(1H1////6X,10H2D RESULTS//)
      END




      SUBROUTINE GRAD (ENEW,RMAX,ALMA,AXIE,SJA,BJA)
C     COMPUTES SPACE OPERATORS FOR 2D PLASMA EQUATIONS
      USE NAME3
      USE NAME10
      USE NAME11
      USE NAME12
      USE NAME13
      DO 10 J=1,3
   10 AA1(J)=0.0
      ENEW=0.0
      RMAX=0.0
      ALMA=0.0
      GAXR=0.0
      GAXZ=0.0
      SJA=-1000.0
      BJA=1000.0
      HUU=HU*HU
      B1=1.0/(2.0*HU*HU)
      B2=1.0/(2.0*HS)
      HS4=4.0*HS
      GAM1=GAM-1.0
C     COMPUTE COEFFICIENTS FOR PLASMA EQUATIONS
      DO 20 J=1,N1
      X1=RB(J)-RA
      X2=ZB(J)-ZA
      AF(J)=X1*X1
      AG(J)=2.0*X1*X2
      AH(J)=X2*X2
      A(J)=X1*X1+X2*X2
      B(J)=RBU(J)*RBU(J)+ZBU(J)*ZBU(J)
      C(J)=X1*RBU(J)+X2*ZBU(J)
   20 H(J)=X1*ZBU(J)-X2*RBU(J)
      DO 30 I=1,NI
      DO 30 J=2,N1
   30 F(I,J)=1.0+EP*(RA+0.5*SL1(I)*(RB(J)-RA)*(R(I,J)+R(I,J+1)))
      DO 40 I=1,NI
      X1=R(I,2)*R(I,2)
```

```
      DO 40 J=2,N1
      X2=R(I,J+1)*R(I,J+1)
      QR(I,J)=X1+X2
   40 X1=X2
      DO 60 I=1,N3
      SUM1=0.0
      DO 50 J=2,N1
      D(I,J)=0.25*H(J)*(QR(I,J)+PS(I)*(QR(I+1,J)-QR(I,J)))
      SUM1=SUM1+F(I,J)*D(I,J)
      SJA=AMAX1(SJA,D(I,J))
   50 BJA=AMIN1(BJA,D(I,J))
      IF (SUM1.LE.0.0) GO TO 200
      PRP(I)=(AM(I)/(SUM1*HU))**GAM
   60 BMP(I)=PRP(I)*SUM1*HU
      DO 80 I=2,NI
      SUM1=0.0
      DO 70 J=2,N1
      DD(I,J)=0.25*H(J)*(QR(I,J)+QS(I-1)*(QR(I,J)-QR(I-1,J)))
      SUM1=SUM1+F(I,J)*DD(I,J)
      SJA=AMAX1(SJA,DD(I,J))
   70 BJA=AMIN1(BJA,DD(I,J))
      IF (SUM1.LE.0.0) GO TO 200
      PRM(I)=(AM(I)/(SUM1*HU))**GAM
   80 BMM(I)=PRM(I)*SUM1*HU
      DO 90 J=2,N1
      G(1,J)=0.5*SA(1)/D(1,J)
      G(NI,J)=0.5*SA(N3)/DD(NI,J)
      DO 90 I=2,N3
   90 G(I,J)=0.5*(SA(I)/D(I,J)+SA(I-1)/DD(I,J))
      DO 100 J=2,N1
      XA(J)=0.0
      XC(J)=0.0
      DO 100 I=1,NI
      X1=(R(I,J+1)-R(I,J))*(R(I,J+1)-R(I,J))/HUU
      X2=0.5*(R(I,J)*R(I,J)+R(I,J+1)*R(I,J+1))
      X3=(R(I,J+1)*R(I,J+1)-R(I,J)*R(I,J))/HU
      X4=(AL(I,J+1)-AL(I,J))*(AL(I,J+1)-AL(I,J))/HUU
      RU2=A(J)*X1+B(J)*X2+C(J)*X3
      QR(I,J)=QQ(I)*RU2/F(I,J)+F(I,J)*X4
      X5=G(I,J)*QQ(I)/F(I,J)
      XA(J)=XA(J)+X1*X5
      XC(J)=XC(J)+0.5*X3*X5
  100 QZ(I,J)=QQ(I)*RU2/(F(I,J)*F(I,J))-X4
      DO 110 J=2,N1
      DO 110 I=1,N3
  110 ENEW=ENEW+0.5*SA(I)*(0.5*(QR(I,J)/D(I,J)+QR(I+1,J)/DD(I+1,J))+(BMP
     1 (I)+BMM(I+1))/GAM1)
      DO 130 J=2,N1
      QR(1,J)=QR(1,J)/(D(1,J)*D(1,J))+2.0*PRP(1)*F(1,J)
      E(1,J)=QZ(1,J)*G(1,J)+SA(1)*PRP(1)*D(1,J)
      XPD(J)=E(1,J)
      XPK(J)=SA(1)*D(1,J)*QR(1,J)
      DO 120 I=2,N3
```

```
      E(I,J)=G(I,J)*QZ(I,J)+SA(I)*PRP(I)*D(I,J)+SA(I-1)*PRM(I)*DD(I,J)
      QZ(I,J)=QR(I,J)/(DD(I,J)*DD(I,J))+2.0*PRM(I)*F(I,J)
      QR(I,J)=QR(I,J)/(D(I,J)*D(I,J))+2.0*PRP(I)*F(I,J)
      XPD(J)=XPD(J)+(1.0-0.5*SL1(I)*(R(I,J)+R(I,J+1)))*E(I,J)
  120 XPK(J)=XPK(J)+SA(I)*QR(I,J)*D(I,J)+SA(I-1)*QZ(I,J)*DD(I,J)
      QZ(NI,J)=QR(NI,J)/(DD(NI,J)*DD(NI,J))+2.0*PRM(NI)*F(NI,J)
      XPK(J)=(XPK(J)+SA(N3)*QZ(NI,J)*DD(NI,J))/(4.0*H(J))
  130 XPD(J)=0.5*XPD(J)
      DO 140 I=1,NI
      G(I,1)=G(I,N1)
      F(I,1)=F(I,N1)
      QR(I,1)=QR(I,N1)
  140 QZ(I,1)=QZ(I,N1)
      DO 150 I=1,N3
  150 E(I,1)=E(I,N1)
      XA(1)=XA(N1)
      XC(1)=XC(N1)
      XPK(1)=XPK(N1)
      XPD(1)=XPD(N1)
      ENEW=ENEW*HS*HU*ZLE
C     COMPUTE MAGNETIC AXIS OPERATORS
      DO 160 J=2,N1
      GAXR=GAXR-EP*XPD(J)+ZBU(J)*XPK(J)-XA(J)*(RB(J)-RA)-XC(J)*RBU(J)
  160 GAXZ=GAXZ-XPK(J)*RBU(J)-XA(J)*(ZB(J)-ZA)-XC(J)*ZBU(J)
      SUMA=0.0
      SUMB=0.0
      SUMC=0.0
C     COMPUTE OPERATOR FOR R EQUATION
      SUM1=0.0
      DO 170 J=2,N1
  170 GR(1,J)=-R(1,J)*SL2(2)*((QR(1,J)-QZ(2,J))*H(J)+(QR(1,J-1)-QZ(2,J-1
     1 ))*H(J-1))/(4.0*HS*PQS(1))
      SUM1=SUM1*HU
      DO 180 J=2,N1
      Y1=(GR(1,J)*PQS(1))/(4.0*R(1,J))
      Y2=(ROA*(AF(J)+AF(J-1))+ROB*(AG(J)+AG(J-1))+ROC*(AH(J)+AH(J-1)))*0
     1 .5
      Y2=Y1/(Y2*Y2)
      SUMA=SUMA+0.5*Y2*(AF(J)+AF(J-1))
      SUMB=SUMB+0.5*Y2*(AG(J)+AG(J-1))
      SUMC=SUMC+0.5*(AH(J)+AH(J-1))*Y2
      Y3=RB(J)+RB(J-1)-2.0*RA
      Y4=ZB(J)+ZB(J-1)-2.0*ZA
      GAXR=GAXR+Y2*(Y3*ROA+Y4*ROB)
      GAXZ=GAXZ+Y2*(Y3*ROB+Y4*ROC)
      DO 180 I=2,N3
      X1=(RB(J)-RA)*E(I,J)+(RB(J-1)-RA)*E(I,J-1)
      X2=G(I,J)*(A(J)*(R(I,J+1)-R(I,J))/HUU+R(I,J)*(C(J)/HU-0.5*B(J)))/F
     1 (I,J)-G(I,J-1)*(A(J-1)*(R(I,J)-R(I,J-1))/HUU+R(I,J)*(C(J-1)/HU+0.
     2 5*B(J-1)))/F(I,J-1)
      X3=(SL2(I+1)-2.0*SL2(I))*(QR(I,J)*H(J)+QR(I,J-1)*H(J-1))+(2.0*SL2
     1 (I)-SL2(I-1))*(QZ(I,J)*H(J)+QZ(I,J-1)*H(J-1))-SL2(I+1)*(QZ(I+1,J)
     2 *H(J)+QZ(I+1,J-1)*H(J-1))+SL2(I-1)*(QR(I-1,J)*H(J)+QR(I-1,J-1)*H
```

```
    3 (J-1))
      GR(I,J)=-(E4*SL1(I)*X1+QQ(I)*X2+(R(I,J)*X3)/(8.0*HS))/PQS(I)
      GR(I,J)=GR(I,J)/(SA1*SL2(I))
      AA=ABS(GR(I,J))
  180 RMAX=AMAX1(RMAX,AA)
C     COMPUTE OPERATOR FOR PSI EQUATION
      DO 190 I=2,NI
      DO 190 J=2,N1
      GZ(I,J)=-(F(I,J)*G(I,J)*(AL(I,J+1)-AL(I,J))-F(I,J-1)*G(I,J-1)*(AL
    1 (I,J)-AL(I,J-1)))/(HUU*PQS(I))
      GZ(I,J)=GZ(I,J)/SA2
      AA=ABS(GZ(I,J))
      ALMA=AMAX1(AA,ALMA)
  190 CONTINUE
      X1=HU/(SL2(2)*ABS(SEAX))
      SUMA=SUMA*X1
      SUMB=SUMB*X1
      SUMC=SUMC*X1
      GAXR=(GAXR*HS*HU)/SA3
      GAXZ=(GAXZ*HS*HU)/SA3
      AA=ABS(GAXR)
      BB=ABS(GAXZ)
      AXIE=AMAX1(AA,BB)
      GO TO 210
  200 PRINT 220
      PRINT 230, ((D(I,J),I=1,N3),J=2,N1)
      CALL ASOUT
      STOP
  210 RETURN
C
  220 FORMAT (1H1,6X,17HNEGATIVE JACOBIAN//)
  230 FORMAT (2X,11F9.3)
      END




      SUBROUTINE ASOR (NCO,ERVA,EVNE,IT)
C     COMPUTES ITERATIONS FOR VACUUM EQUATIONS
      USE NAME3
      USE NAME11
      USE NAME12
      USE NAME13
      A22=0.0
      PM1=1.0-OM1
      AA1(4)=0.0
      DSU=1.0/(2.0*HR*HU)
      AT=0.01
      IF (IT.GT.0) AT=0.1
```

```
      IT=1
C     COMPUTE COEFFICIENTS FOR VACUUM EQUATIONS
      DO 20 J=2,N1
      X7=RR(J)-RB(J)
      X8=ZZ(J)-ZB(J)
      DO 10 I=1,N6
      S=(I-0.5)*HR
      S1=1.0-S
      Y1=S1*RBU(J)+S*RU(J)
      Y2=S1*ZBU(J)+S*ZU(J)
      AK=1.0+EP*(RB(J)+S*X7)
      DEL=(X7*Y2-Y1*X8)/AK
      D(I,J)=DEL
      E(I,J)=(Y1*Y1+Y2*Y2)/DEL
      F(I,J)=(X7*X7+X8*X8)/DEL
   10 G(I,J)=-(Y1*X7+Y2*X8)/DEL
   20 CONTINUE
      DO 30 J=2,N1
      DO 30 I=1,N6
   30 A22=A22+D(I,J)
      A22=(A22*HR*HU)/ZLE
      DO 40 I=1,N6
      E(I,1)=E(I,N1)
      F(I,1)=F(I,N1)
   40 G(I,1)=G(I,N1)
   50 NCO=NCO+1
      ERVA=0.0
      AA3(4)=AA1(4)
      AA1(4)=0.0
C     SUCCESSIVE OVERRELAXATION
      DO 70 J=2,N1
      SUM1=H1*(E(1,J)+E(1,J-1))*PA(2,J)+H2*(F(1,J)*PA(1,J+1)+F(1,J-1)*PA
     1 (1,J-1))+H3*(G(1,J)*PA(2,J+1)-G(1,J-1)*PA(2,J-1))
      SUM2=H1*(E(1,J)+E(1,J-1))+H2*(F(1,J)+F(1,J-1))+H3*(G(1,J)-G(1,J-1)
     1 )
      X1=SUM1-PA(1,J)*SUM2
      AA1(4)=AA1(4)+X1*X1
      PAS=PM1*PA(1,J)+(OM1*SUM1)/SUM2
      AA=ABS(X1)
      ERVA=AMAX1(ERVA,AA)
      PA(1,J)=PAS
      DO 60 I=2,N6
      SUM1=H1*((E(I,J)+E(I,J-1))*PA(I+1,J)+(E(I-1,J)+E(I-1,J-1))*PA(I-1
     1 ,J))+H2*((F(I,J)+F(I-1,J))*PA(I,J+1)+(F(I,J-1)+F(I-1,J-1))*PA(I,J
     2 -1))+H3*(G(I,J)*PA(I+1,J+1)-G(I,J-1)*PA(I+1,J-1)-G(I-1,J)*PA(I-1
     3 ,J+1)+G(I-1,J-1)*PA(I-1,J-1))
      SUM2=H1*(E(I,J)+E(I,J-1)+E(I-1,J)+E(I-1,J-1))+H2*(F(I,J)+F(I-1,J)
     1 +F(I,J-1)+F(I-1,J-1))+H3*(G(I,J)-G(I,J-1)-G(I-1,J)+G(I-1,J-1))
      X1=SUM1-PA(I,J)*SUM2
      AA1(4)=AA1(4)+X1*X1
      PAS=PM1*PA(I,J)+(OM1*SUM1)/SUM2
      AA=ABS(X1)
      ERVA=AMAX1(ERVA,AA)
```

```
   60 PA(I,J)=PAS
   70 CONTINUE
      DO 80 I=1,N6
      PA(I,1)=PA(I,N1)-C1
   80 PA(I,N2)=PA(I,2)+C1
      IF (NCO.GT.300) GO TO 110
      IF (ERVA.GT.AT) GO TO 50
      IF (NCO.LT.NV) GO TO 50
      A11=0.0
      DO 100 J=2,N1
      DO 90 I=1,N6
      X1=PA(I+1,J)-PA(I,J)
      X2=PA(I+1,J+1)-PA(I,J)
      X3=PA(I,J+1)-PA(I+1,J)
      X4=PA(I+1,J+1)-PA(I,J+1)
      X5=PA(I,J+1)-PA(I,J)
      X6=PA(I+1,J+1)-PA(I+1,J)
   90 A11=A11+0.5*(H1*E(I,J)*(X1*X1+X4*X4)+H2*F(I,J)*(X5*X5+X6*X6)+H3*G
     1 (I,J)*(X2*X2-X3*X3))
  100 CONTINUE
      A11=ZLE*HR*HU*A11
      EVNE=0.5*(A11+C2*A22*C2)
      GO TO 120
  110 PRINT 130
      CALL ASOUT
      STOP
  120 CONTINUE
      AA1(4)=AA1(4)*HU*HR
      AA3(4)=AA3(4)*HU*HR
      AA2(4)=0.5*AA1(4)
      RETURN
C
  130 FORMAT (///,6X,34HLAPLACE EQUATION DOES NOT CONVERGE)
      END




      SUBROUTINE ASBO (ERBO,ERBO1)
C     COMPUTES ITERATION OF 2D FREE BOUNDARY EQUATION
      USE NAME3
      USE NAME10
      USE NAME12
      USE NAME13
      C11=1.0
      C22=C2*C2
      Z22=ZLE*ZLE
      ERBO=0.0
      SUM1=0.0
```

```
      SUM2=0.0
      HS8=8.0*HS
      PB=PRM(NI)
      Q2=Q(NI)*Q(NI)
C     COMPUTE COEFFICIENTS FOR LAX-WENDROFF METHOD
      DO 10 J=2,N1
      GU2=(PA(1,J+1)-PA(1,J))*(PA(1,J+1)-PA(1,J))*H2
      AU=RBU(J)
      BU=ZBU(J)
      AK=1.0+EP*RB(J)
      ABU=AU*AU+BU*BU
      PL=0.25*H(J)*(2.0+QS(N3)*(2.0-(R(N3,J)*R(N3,J)+R(N3,J+1)*R(N3,J+1)
     1 )))
      X1=RR(J)-R2
      X2=ZZ(J)-Z2
      PU2=(AL(NI,J+1)-AL(NI,J))*(AL(NI,J+1)-AL(NI,J))*H2
      PAS1=C11*GU2/ABU
      BV2=PAS1+C22/(Z22*AK*AK)
      BP2=(AK*AK*PU2+Q2*ABU)/(AK*AK*PL*PL)
      AK2(J)=AK*(0.5*(BP2-BV2)+PB)
      X3=RU(J)
      X4=ZU(J)
      PAS2=AU*X3+BU*X4
      PAS3=Q2/(PL*PL*AK*AK)
      PAS4=(X(J)+X(J+1))*(X1*X4-X2*X3)+(R2-RA)*X4-(Z2-ZA)*X3
      R30=RB(J)-RA
      Z30=ZB(J)-ZA
      PAS5=BP2/(R30*BU-Z30*AU)
      PAS6=PAS3+PAS1/ABU
      AKF(J)=EP*X1*(AK2(J)/AK-PAS3*ABU+C22/(Z22*AK*AK))+AK*(PAS6*PAS2
     1 -PAS5*PAS4)
   10 AKFU(J)=AK*(PAS6*(X1*AU+X2*BU)-PAS5*((R2-RA)*X2-(Z2-ZA)*X1))
      AK2(1)=AK2(N1)
      AK2(N2)=AK2(2)
      AKF(1)=AKF(N1)
      AKFU(1)=AKFU(N1)
C     COMPUTE LAX-WENDROFF ITERATION
      DO 20 J=2,N1
      X1=0.5*(AK2(J)+AK2(J-1))
      X2=0.5*(AKF(J)+AKF(J-1))
      X3=0.5*(AKFU(J)+AKFU(J-1))
      X4=(AK2(J)-AK2(J-1))/HU
      X5=X1+0.5*DTB*(X1*X2+X3*X4)
      SUM1=SUM1+0.5*X5*(ZBU(J)+ZBU(J-1))*(1.0-X(J))
      SUM2=SUM2-0.5*X5*(RBU(J)+RBU(J-1))*(1.0-X(J))
      AA=ABS(X5)
      ERBO=AMAX1(ERBO,AA)
      X(J)=X(J)+DTB*X5
   20 CONTINUE
      SUM1=SUM1*HU
      SUM2=SUM2*HU
      X(1)=X(N1)
      X(N2)=X(2)
```

```
C       COMPUTE VACUUM AXIS ITERATION
        RETURN
        END




        SUBROUTINE EXTEN
C       EXTENSION OF EQUILIBRIUM SOLUTION TO SEVERAL PERIODS FOR STABILITY
C       ANALYSIS
        USE NAME1
        USE NAME2
        USE NAME6
        USE NAME7
        USE NAME8
        USE NAME9
        USE NAME10
        USE NAME11
        USE NAME12
        USE NAME13
        USE NAME14
        USE NAME15
        USE NAME16
        USE NAME18
        USE NAME20
        USE NAME21
        NKO=NK
        NK1=NKO+1
        NN=NRUN-1
        NK=NK*NRUN
        N4=NK+1
        N5=NK+2
        HV=1.0/NK
        ZLE=ZLE*NRUN
        A3=1.0/(4.0*HU*HU*ZLE*ZLE)
        A4=1.0/(4.0*HV*HV*ZLE*ZLE)
        A5=1.0/(4.0*HU*HV*ZLE*ZLE)
        HV4=1.0/(4.0*HV*HV)
        HUV=1.0/(4.0*HU*HV)
        DO 10 K=1,N5
        V=2.0*PI*(K-2.5)*HV
        VP1=2.0*PI*(K-2.0)*HV
        SVP(1,K)=1.0
        SVP(2,K)=SIN(VP1)
        SVP(3,K)=COS(VP1)
        SVP(4,K)=SIN(2.0*VP1)
        SVP(5,K)=COS(2.0*VP1)
        SVP(6,K)=SIN(3.0*VP1)
        SVP(7,K)=COS(3.0*VP1)
```

```
        SV(1,K)=1.0
        SV(2,K)=SIN(V)
        SV(3,K)=COS(V)
        SV(4,K)=SIN(2.0*V)
        SV(5,K)=COS(2.0*V)
        SV(6,K)=SIN(3.0*V)
   10   SV(7,K)=COS(3.0*V)
        DO 20 I=1,NI
        AM(I)=AM(I)*NRUN
        AMS(I)=AMS(I)*NRUN
        PC(I)=PC(I)*NRUN
   20   Q(I)=Q(I)*NRUN
        DO 30 K=2,NK1
        DO 30 KK=1,NN
        K1=K+NKO*KK
        RA(K1)=RA(K)
        RN(K1)=RN(K)
        ZA(K1)=ZA(K)
        ZN(K1)=ZN(K)
        ROAX(K1)=ROAX(K)
        ROBX(K1)=ROBX(K)
   30   ROCX(K1)=ROCX(K)
        RA(1)=RA(N4)
        RA(N5)=RA(2)
        ZA(1)=ZA(N4)
        ZA(N5)=ZA(2)
        RN(1)=RN(N4)
        RN(N5)=RN(2)
        ZN(1)=ZN(N4)
        ZN(N5)=ZN(2)
        ROAX(1)=ROAX(N4)
        ROAX(N5)=ROAX(2)
        ROBX(1)=ROBX(N4)
        ROBX(N5)=ROBX(2)
        ROCX(1)=ROCX(N4)
        ROCX(N5)=ROCX(2)
        DO 50 I=1,NI
        DO 50 J=1,N2
        DO 40 KK=1,NN
        DO 40 K=2,NK1
        K1=K+KK*NKO
        RO(I,J,K1)=RO(I,J,K)
        V=(K1-2.5)*HV
   40   AL(I,J,K1)=AL(I,J,K)+(Q(I)*KK)/NRUN
        RO(I,J,1)=RO(I,J,N4)
        RO(I,J,N5)=RO(I,J,2)
        AL(I,J,1)=AL(I,J,N4)-Q(I)
   50   AL(I,J,N5)=AL(I,J,2)+Q(I)
        DO 80 J=1,N2
        DO 60 K=2,NK1
        RV(J,K)=RV(J,K)*NRUN
   60   ZV(J,K)=ZV(J,K)*NRUN
        DO 70 KK=1,NN
```

```
      DO 70 K=2,NK1
      K1=K+KK*NKO
      R(J,K1)=R(J,K)
      Z(J,K1)=Z(J,K)
      RU(J,K1)=RU(J,K)
      RV(J,K1)=RV(J,K)
      ZU(J,K1)=ZU(J,K)
   70 ZV(J,K1)=ZV(J,K)
      R(J,1)=R(J,N4)
      R(J,N5)=R(J,2)
      Z(J,1)=Z(J,N4)
      Z(J,N5)=Z(J,2)
      RU(J,1)=RU(J,N4)
      RU(J,N5)=RU(J,2)
      RV(J,1)=RV(J,N4)
      RV(J,N5)=RV(J,2)
      ZU(J,1)=ZU(J,N4)
      ZU(J,N5)=ZU(J,2)
      ZV(J,1)=ZV(J,N4)
   80 ZV(J,N5)=ZV(J,2)
      ETOT=ETOT*NRUN
      ENER=ENER*NRUN
      EVNE=EVNE*NRUN
      DO 90 K=2,NK1
      DO 90 KK=1,NN
      K1=K+NKO*KK
      RVA(K1)=RVA(K)
   90 ZVA(K1)=ZVA(K)
      RVA(1)=RVA(N4)
      RVA(N5)=RVA(2)
      ZVA(1)=ZVA(N4)
      ZVA(N5)=ZVA(1)
      DO 110 J=1,N2
      DO 100 KK=1,NN
      DO 100 K=2,NK1
      K1=K+KK*NKO
  100 X(J,K1)=X(J,K)
      X(J,1)=X(J,N4)
  110 X(J,N5)=X(J,2)
      IF (NVAC.LT.0) GO TO 170
      C2=C2*NRUN
      DO 160 I=1,NIV
      DO 160 J=1,N2
      DO 150 KK=1,NN
      DO 150 K=2,NK1
      K1=K+KK*NKO
  150 PT(I,J,K1)=PT(I,J,K)+(C2*KK)/NRUN
      PT(I,J,1)=PT(I,J,N4)-C2
  160 PT(I,J,N5)=PT(I,J,2)+C2
  170 CONTINUE
      RETURN
      END
```

```
      SUBROUTINE PROJ(X1,ITER)
C     PROJECTS SOLUTION ONTO LINEAR SUBSPACE TO INITIALIZE
C     STABILITY CALCULATION
      USE NAME1
      USE NAME2
      USE NAME6
      USE NAME7
      USE NAME8
      USE NAME9
      USE NAME10
      USE NAME11
      USE NAME12
      USE NAME13
      USE NAME14
      USE NAME15
      USE NAME16
      USE NAME18
      USE NAME20
      USE NAME21
      DIMENSION AA(5,5), BB(5), CC(5), AW(5,5), WKS1(5), WKS2(5)
      DO 10 K=1,N5
      RA(K)=RA(K)+X1*(EL1*EH1(K)+EL2*EH2(K))
      ZA(K)=ZA(K)+X1*(EM1*EJ1(K)+EM2*EJ2(K))
      RN(K)=RA(K)
      ZN(K)=ZA(K)
      DO 10 J=1,N2
      DO 10 I=1,NI
      RO(I,J,K)=RO(I,J,K)+X1*(EAM1(I)*EF1(J,K)+EAM2(I)*EF2(J,K)+EAM3(I)
     1 *EF3(J,K))
      AL(I,J,K)=AL(I,J,K)-X1*(EAM4(I)*EG1(J,K)+EAM5(I)*EG2(J,K)+EAM6(I)
     1 *EG3(J,K))
      XO(I,J,K)=RO(I,J,K)
   10 XL(I,J,K)=AL(I,J,K)
      IF (NVAC.LT.0) GO TO 30
      DO 20 J=1,N2
      DO 20 K=1,N5
   20 X(J,K)=X(J,K)+X1*EBD(J,K)
   30 CONTINUE
      NAC=ITER+NAC
C     INITIALIZE RO AXIS
      DO 60 K=2,N4
      DO 40 J=1,3
      BB(J)=0.0
      DO 40 JJ=1,3
   40 AA(J,JJ)=0.0
      DO 50 J=2,N1
      X4=0.25*(R(J,K)+R(J,K-1)+R(J-1,K)+R(J-1,K-1))-RA(K)
      X5=0.25*(Z(J,K)+Z(J-1,K)+Z(J,K-1)+Z(J-1,K-1))-ZA(K)
      X1=X4*X4+X5*X5
      X1=X1*RBOU*RBOU
      Y1=1.0/(RO(1,J,K)*RO(1,J,K))
      XX1=X4*X4
      XX2=2.0*X4*X5
```

```
      XX3=X5*X5
      BB(1)=BB(1)+Y1*XX1
      BB(2)=BB(2)+Y1*XX2
      BB(3)=BB(3)+Y1*XX3
      AA(1,1)=AA(1,1)+XX1*XX1
      AA(1,2)=AA(1,2)+XX1*XX2
      AA(1,3)=AA(1,3)+XX1*XX3
      AA(2,2)=AA(2,2)+XX2*XX2
      AA(3,3)=AA(3,3)+XX3*XX3
 50   AA(2,3)=AA(2,3)+XX3*XX2
      AA(2,1)=AA(1,2)
      AA(3,1)=AA(1,3)
      AA(3,2)=AA(2,3)
      CALL F04ATF (AA,5,BB,3,CC,AW,5,WKS1,WKS2,IFAIL)
      ROAX(K)=CC(1)
      ROBX(K)=CC(2)
      ROCX(K)=CC(3)
 60   CONTINUE
      CALL CBO (N1)
      DO 100 J=2,N1
      DO 70 K=1,N4
      RB2(K)=RB1(K)
 70   ZB2(K)=ZB1(K)
      CALL CBO (J)
      DO 80 K=1,N4
      D1(K)=RB1(K)-0.5*(RA(K)+RA(K+1))
      D2(K)=ZB1(K)-0.5*(ZA(K)+ZA(K+1))
      D3(K)=RB2(K)-0.5*(RA(K)+RA(K+1))
      D4(K)=ZB2(K)-0.5*(ZA(K)+ZA(K+1))
      CAF1(K)=D1(K)*D1(K)
      CAG1(K)=2.0*D1(K)*D2(K)
      CAH1(K)=D2(K)*D2(K)
      CAF2(K)=D3(K)*D3(K)
      CAG2(K)=2.0*D3(K)*D4(K)
      CAH2(K)=D4(K)*D4(K)
 80   CONTINUE
      DO 90 K=2,N4
      X1=0.25*(CAF1(K)+CAF1(K-1)+CAF2(K)+CAF2(K-1))
      X2=0.25*(CAG1(K)+CAG1(K-1)+CAG2(K)+CAG2(K-1))
      X3=0.25*(CAH1(K)+CAH1(K-1)+CAH2(K)+CAH2(K-1))
      Y1=ROAX(K)*X1+ROBX(K)*X2+ROCX(K)*X3
      RO(1,J,K)=1.0/SQRT(Y1)
 90   CONTINUE
 100  CONTINUE
      DO 110 J=2,N1
      RO(1,J,1)=RO(1,J,N4)
 110  RO(1,J,N5)=RO(1,J,2)
      DO 120 K=1,N5
      RO(1,1,K)=RO(1,N1,K)
 120  RO(1,N2,K)=RO(1,2,K)
      DO 130 J=1,N2
      DO 130 K=1,N5
 130  XO(1,J,K)=RO(1,J,K)
```

```
      RETURN
      END




      SUBROUTINE TGRAD (BJA,SJA,EAX,ERO,EAL,ERBO,EROAX)
C     ITERATION TO SOLVE THE   PARTIAL DIFFERENTIAL EQUATIONS FOR
C     THE PLASMA IN 3D, WHICH IS THE MOST IMPORTANT PART OF THE
C     CALCULATION
      USE NAME1
      USE NAME2
      USE NAME6
      USE NAME7
      USE NAME8
      USE NAME10
      USE NAME11
      USE NAME12
      USE NAME13
      USE NAME14
      USE NAME20
C     INITIALIZE VARIABLES AND COMPUTE CONSTANTS
      DO 10 J=1,3
   10 AA1(J)=0.0
      DO 20 I=1,NI
      SPRM(I)=0.0
   20 SPR(I)=0.0
      ENER=0.0
      RBO2=RBOU*RBOU
      RBO4=RBO2*RBO2
      PRQ=0.0
      PRQ1=0.0
      BJA=1000.0
      SJA=-1000.0
      CALL CBO (N1)
      CALL CIN (1,BJA,SJA)
      BJA=1000.0
      SJA=-1000.0
      DO 30 I=1,NI
      BTP(I)=0.0
      VOL(I)=0.0
      SPVOL(I)=0.0
      SPM(I)=0.0
      SPPVOL(I)=0.0
      BPP(I)=0.0
      BET(I)=0.0
      PC(I)=0.0
   30 TC(I)=0.0
      FAC1=HU*HV
```

```
      FAC=FAC1/ZLE
      ENER=0.0
      DO 40 I=1,NI
      SPRM(I)=0.0
   40 SPR(I)=0.0
      ERBO=0.0000001
      EAX=0.0
      ERO=0.0
      EROAX=0.0
      EAL=0.0
      HU2=2.0*HU
      HV2=2.0*HV
      HUV2=2.0*HU*HV
      HUU2=2.0*HU*HU
      HVV2=2.0*HV*HV
      HVV=2.0/(HV*HV)
      HV1=1.0/HV
      HSU=HS*HU
      FRO=SL2(2)/(8.0*HS*PQS(1))
      DO 50 K=2,N4
      SUMA(K)=0.0
      SUMB(K)=0.0
      SUMC(K)=0.0
      SUMRO(K)=0.0
      SUMR(K)=0.0
   50 SUMZ(K)=0.0
      DO 320 J=2,N1
C     SOLVE PDE ON EACH U=CONSTANT
C     COEFFICIENTS IN J+1/2 BECOME COEFFICIENTS IN J-1/2 FOR NEW J
      DO 60 K=1,N4
      RB2(K)=RB1(K)
      ZB2(K)=ZB1(K)
      RBU2(K)=RBU1(K)
      ZBU2(K)=ZBU1(K)
      RBV2(K)=RBV1(K)
      ZBV2(K)=ZBV1(K)
      HB2(K)=HB1(K)
      CA2(K)=CA1(K)
      CAF2(K)=CAF1(K)
      CAG2(K)=CAG1(K)
      CAH2(K)=CAH1(K)
      CB2(K)=CB1(K)
      CC2(K)=CC1(K)
      CE2(K)=CE1(K)
      CF2(K)=CF1(K)
      CG2(K)=CG1(K)
      CL2(K)=CL1(K)
      CM2(K)=CM1(K)
      CN2(K)=CN1(K)
      XA2(K)=XA1(K)
      XB2(K)=XB1(K)
      XC2(K)=XC1(K)
      XG2(K)=XG1(K)
```

```
      XN2(K)=XN1(K)
      XD2(K)=XD1(K)
      XF2(K)=XF1(K)
      XP2(K)=XP1(K)
   60 XQ2(K)=XQ1(K)
      DO 80 I=2,NI
      DO 70 K=1,N4
   70 PM2(I,K)=PM1(I,K)
   80 CONTINUE
      DO 100 I=1,NI
      DO 90 K=1,N4
      U2(I,K)=U1(I,K)
      V2(I,K)=V1(I,K)
      UV2(I,K)=UV1(I,K)
      E2(I,K)=E1(I,K)
      F2(I,K)=F1(I,K)
      G2(I,K)=G1(I,K)
      P2(I,K)=P1(I,K)
   90 Q2(I,K)=Q1(I,K)
  100 CONTINUE
C     COMPUTE COEFFICIENTS AT J+1/2
      CALL CBO (J)
      CALL CIN (J,BJA,SJA)
      IF (NPLOT.LT.0) GO TO 130
      X5=ZLE*HU*HV
      PR(1)=PRP(1)
      PR(NI)=PRM(NI)
      DO 110 I=2,N3
      PR(I)=0.5*(PRP(I)+PRM(I))
      X1=2.0*FAC1*PR(I)
      DO 110 K=2,N4
      SPM(I)=SPM(I)+X5*BK(I,K)*DM(I,K)
      SPVOL(I)=SPVOL(I)+X5*BK(I,K)*D(I,K)
  110 BET(I)=BET(I)+X1*BK(I,K)/(P1(I,K)+PM1(I,K))
      DO 120 K=2,N4
      SPVOL(1)=SPVOL(1)+X5*BK(1,K)*D(1,K)
      SPM(NI)=SPM(NI)+X5*BK(NI,K)*DM(NI,K)
      BET(1)=BET(1)+FAC1*PR(1)*BK(1,K)/P1(1,K)
  120 BET(NI)=BET(NI)+FAC1*PR(NI)*BK(NI,K)/PM1(NI,K)
C     COMPUTE CURRENTS
  130 DO 150 I=1,NI
      X1=0.5*FAC/PQS(I)
      DO 140 K=2,N4
      PU=(AL(I,J+1,K)+AL(I,J+1,K+1)-AL(I,J,K)-AL(I,J,K+1))/(2.0*HU)
      PV=(AL(I,J,K+1)+AL(I,J+1,K+1)-AL(I,J,K)-AL(I,J+1,K))/(2.0*HV)
      BU=PV*E1(I,K)-PU*F1(I,K)
      BV=PV*F1(I,K)-PU*G1(I,K)
      PC(I)=PC(I)-BV*X1
      TC(I)=TC(I)-BU*X1
      IF (NPLOT.LT.0) GO TO 140
      RU2=E1(I,K)*BK(I,K)/DD(I,K)+0.0000001
      RV2=G1(I,K)*BK(I,K)/DD(I,K)
      BPP(I)=BPP(I)-X1*BU/SQRT(RU2)
```

```
      BTP(I)=BTP(I)-X1*BV/SQRT(RV2)
140 CONTINUE
150 CONTINUE
      IF (NPLOT.GT.0) GO TO 320
C     ITERATION FOR PSI EQUATION
      DO 160 K=1,N4
160 XL(1,J,K)=G1(1,K)
      DO 190 I=2,NI
      X4=OP1(I)*SL2(I)*OP2(I)/SA2
      X3=X4*PQS(I)
      X5=X3*SA2
      DO 170 K=2,N4
      D1(K)=(E1(I,K)+E2(I,K))*(AL(I,J,K+1)-AL(I,J,K))-(E1(I,K-1)+E2(I,K-
    1 1))*(AL(I,J,K)-AL(I,J,K-1))
      D2(K)=(G1(I,K)+G1(I,K-1))*(AL(I,J+1,K)-AL(I,J,K))-(G2(I,K)+G2(I,K-
    1 1))*(AL(I,J,K)-XL(I,J-1,K))
170 D3(K)=F1(I,K)*(AL(I,J+1,K+1)-AL(I,J,K))+F2(I,K)*(AL(I,J,K)-XL(I,J-
    1 1,K+1))-F1(I,K-1)*(AL(I,J+1,K-1)-AL(I,J,K))-F2(I,K-1)*(AL(I,J,K)
    2 -XL(I,J-1,K-1))
      DO 180 K=2,N4
      DAL=-(A4*D1(K)+A3*D2(K)-A5*D3(K))/PQS(I)
      DAL=DAL/SA2
      X2=EAM4(I)*EG1(J,K)+EAM5(I)*EG2(J,K)+EAM6(I)*EG3(J,K)
      X2=-X2
      PRQ=PRQ+X5*X2*(DAL-RA2*(AL(I,J,K)-XL(I,J,K)))
      DAL1=DAL-RQ*X2*X4
      AA=ABS(DAL1)
      EAL=AMAX1(EAL,AA)
      PAS=DG2*(AL(I,J,K)*DH2-DAL-RA2*XL(I,J,K))
      XL(I,J,K)=AL(I,J,K)
180 AL(I,J,K)=PAS
      AL(I,J,1)=AL(I,J,N4)-Q(I)
      AL(I,J,N5)=AL(I,J,2)+Q(I)
      XL(I,J,1)=XL(I,J,N4)-Q(I)
190 XL(I,J,N5)=XL(I,J,2)+Q(I)
C     ITERATION FOR R EQUATION
      DO 200 K=2,N4
      DRO=-FRO*RO(1,J,K)*(HB1(K)*(P1(1,K)-PM1(2,K))+HB2(K)*(P2(1,K)-PM2(
    1 2,K))+HB1(K-1)*(P1(1,K-1)-PM1(2,K-1))+HB2(K-1)*(P2(1,K-1)-PM2(2,K
    2 -1)))
      DRO=(DRO*PQS(1))/(2.0*RO(1,J,K))
      X4=0.25*(RB1(K)+RB1(K-1)+RB2(K)+RB2(K-1))-RA(K)
      X5=0.25*(ZB1(K)+ZB1(K-1)+ZB2(K)+ZB2(K-1))-ZA(K)
      X1=0.25*(CAF1(K)+CAF1(K-1)+CAF2(K)+CAF2(K-1))
      X2=0.25*(CAG1(K)+CAG1(K-1)+CAG2(K)+CAG2(K-1))
      X3=0.25*(CAH1(K)+CAH1(K-1)+CAH2(K)+CAH2(K-1))
      Y1=ROAX(K)*X1+ROBX(K)*X2+ROCX(K)*X3
      XO(1,J,K)=RO(1,J,K)
      RO(1,J,K)=SAFAX/SQRT(Y1)+(1.0-SAFAX)*RO(1,J,K)
      Y1=DRO/(Y1*Y1)
      SUMA(K)=SUMA(K)+Y1*X1
      SUMB(K)=SUMB(K)+Y1*X2
      SUMC(K)=SUMC(K)+Y1*X3
```

```
      Y1=2.0*Y1
      SUMR(K)=SUMR(K)+Y1*(ROAX(K)*X4+ROBX(K)*X5)
      SUMZ(K)=SUMZ(K)+Y1*(ROBX(K)*X4+ROCX(K)*X5)
  200 CONTINUE
      RO(1,J,1)=RO(1,J,N4)
      RO(1,J,N5)=RO(1,J,2)
      XO(1,J,1)=XO(1,J,N4)
      XO(1,J,N5)=XO(1,J,2)
      DO 290 I=2,N3
      XRQ=PQS(I)*OP1(I)*SL2(I)
      XRQ1=RQ*OP1(I)*SL2(I)
      X1=A1*SL1(I)
      X2=SL2(I)/HUU2
      X3=SL2(I)/HVV2
      X4=0.25*SL2(I)
      X5=SL2(I)/HU2
      X6=SL2(I)/HV2
      X7=SL2(I)/HUV2
      X8=0.25*SL1(I)
      X9=SL1(I)/HV2
      X10=SL1(I)/HU2
      X11=2.0*SL1(I)
      DO 210 K=2,N4
      D1(K)=Q1(I,K)*(RB1(K)-0.5*(RA(K)+RA(K+1)))+Q2(I,K)*(RB2(K)-0.5*(RA
     1 (K)+RA(K+1)))
  210 D2(K)=HB1(K)*(DP(I)*P1(I,K)+DQ(I-1)*PM1(I,K)-XS(I+1)*PM1(I+1,K)+XS
     1 (I-1)*P1(I-1,K))+HB2(K)*(DP(I)*P2(I,K)+DQ(I-1)*PM2(I,K)-XS(I+1)
     2 *PM2(I+1,K)+XS(I-1)*P2(I-1,K))
      DO 220 K=2,N4
      D3(K)=(CA1(K)*V1(I,K)+CA1(K-1)*V1(I,K-1))*(RO(I,J+1,K)-RO(I,J,K))-
     1 (CA2(K)*V2(I,K)+CA2(K-1)*V2(I,K-1))*(RO(I,J,K)-XO(I,J-1,K))
      D4(K)=(CA1(K)*U1(I,K)+CA2(K)*U2(I,K))*(RO(I,J,K+1)-RO(I,J,K))
  220 D5(K)=CB1(K)*V1(I,K)+CE1(K)*U1(I,K)-2.0*CM1(K)*UV1(I,K)+CB2(K)*V2
     1 (I,K)+CE2(K)*U2(I,K)-2.0*CM2(K)*UV2(I,K)
      DO 230 K=2,N4
      D6(K)=CC1(K)*V1(I,K)-CG1(K)*UV1(I,K)-(CC2(K)*V2(I,K)-CG2(K)*UV2(I
     1 ,K))
  230 D7(K)=CG1(K)*U1(I,K)-CC1(K)*UV1(I,K)+CG2(K)*U2(I,K)-CC2(K)*UV2(I,K
     1 )
      D1(1)=D1(N4)
      D2(1)=D2(N4)
      D4(1)=D4(N4)
      D5(1)=D5(N4)
      D6(1)=D6(N4)
      D7(1)=D7(N4)
      DO 240 K=2,N4
  240 D8(K)=-X1*(D1(K)+D1(K-1))-X2*D3(K)-X3*(D4(K)-D4(K-1))+RO(I,J,K)*
     1 (X4*(D5(K)+D5(K-1))-X5*(D6(K)+D6(K-1))-X6*(D7(K)-D7(K-1))-0.125*
     2 (D2(K)+D2(K-1)))
      DO 250 K=2,N4
      D1(K)=CL1(K)*U1(I,K)-CN1(K)*UV1(I,K)+CL2(K)*U2(I,K)-CN2(K)*UV2(I,K
     1 )
      D2(K)=CD1(K)*(U1(I,K)+U2(I,K))
```

```
      D3(K)=CF1(K)*U1(I,K)+CF2(K)*U2(I,K)
      D4(K)=CF1(K)*UV1(I,K)-CF2(K)*UV2(I,K)
  250 D5(K)=CA1(K)*UV1(I,K)*(RO(I,J+1,K+1)-RO(I,J,K))+CA2(K)*UV2(I,K)*
    1 (RO(I,J,K)-XO(I,J-1,K+1))
      D1(1)=D1(N4)
      D2(1)=D2(N4)
      D3(1)=D3(N4)
      D4(1)=D4(N4)
      D5(1)=D5(N4)
      DO 260 K=2,N4
  260 D7(K)=D5(K)-CA1(K-1)*UV1(I,K-1)*(RO(I,J+1,K-1)-RO(I,J,K))-CA2(K-1)
    1 *UV2(I,K-1)*(RO(I,J,K)-XO(I,J-1,K-1))
      DO 270 K=2,N4
  270 D6(K)=D8(K)+X8*(1.0-X11*RO(I,J,K))*(D1(K)+D1(K-1))-(1.0-SL1(I)*RO
    1 (I,J,K))*(X8*(D2(K)+D2(K-1))+X9*(D3(K)-D3(K-1))-X10*(D4(K)+D4(K-1
    2 )))+X7*D7(K)
      DO 280 K=2,N4
      DRO=D6(K)/PQS(I)
      YRQ=EAM1(I)*EF1(J,K)+EAM2(I)*EF2(J,K)+EAM3(I)*EF3(J,K)
      YRQ=YRQ*RBO4
      DRO1=(DRO-XRQ1*YRQ)/(SA1*SL2(I))
      AA=ABS(DRO1)
      DRO=DRO1/(SA1*SL2(I))
      PRQ=PRQ+YRQ*XRQ*(DRO-RA1*(RO(I,J,K)-XO(I,J,K)))
      ERO=AMAX1(ERO,AA)
      PAS=DG1*(RO(I,J,K)*DH1-DRO-RA1*XO(I,J,K))
      XO(I,J,K)=RO(I,J,K)
  280 RO(I,J,K)=PAS
      RO(I,J,1)=RO(I,J,N4)
      RO(I,J,N5)=RO(I,J,2)
      XO(I,J,1)=XO(I,J,N4)
  290 XO(I,J,N5)=XO(I,J,2)
C     ITERATION FOR MAGNETIC AXIS EQUATION
      DO 300 K=2,N4
      D1(K)=-XP1(K)-XA1(K)*(RB1(K)-0.5*(RA(K)+RA(K+1)))+0.5*(XQ1(K)*ZBU1
    1 (K)-RBU1(K)*XC1(K)-HV1*(RA(K+1)-RA(K))*XF1(K)-RBV1(K)*XG1(K))
      D2(K)=-HVV*XD1(K)*(RA(K+1)-RA(K))-HV1*(XF1(K)*(RB1(K)-0.5*(RA(K)
    1 +RA(K+1)))+RBV1(K)*XB1(K)+RBU1(K)*XN1(K))
      D3(K)=-0.5*(XQ1(K)*RBU1(K)+XC1(K)*ZBU1(K)+XG1(K)*ZBV1(K)+HV1*XF1(K
    1 )*(ZA(K+1)-ZA(K)))-XA1(K)*(ZB1(K)-0.5*(ZA(K)+ZA(K+1)))
      D4(K)=-HVV*XD1(K)*(ZA(K+1)-ZA(K))-HV1*(XF1(K)*(ZB1(K)-0.5*(ZA(K)
    1 +ZA(K+1)))+ZBV1(K)*XB1(K)+ZBU1(K)*XN1(K))
  300 CONTINUE
      D1(1)=D1(N4)
      D2(1)=D2(N4)
      D3(1)=D3(N4)
      D4(1)=D4(N4)
      DO 310 K=2,N4
      SUMR(K)=SUMR(K)+D1(K)+D1(K-1)+D2(K)-D2(K-1)
  310 SUMZ(K)=SUMZ(K)+D3(K)+D3(K-1)+D4(K)-D4(K-1)
  320 CONTINUE
      IF (NPLOT.GT.0) GO TO 560
      DO 350 K=2,N4
```

```
      SUM=0.0
      AL(1,2,K)=0.0
      DO 330 J=2,N1
  330 SUM=SUM+1.0/(XL(1,J,K)+XL(1,J,K-1))
      DO 340 J=2,N1
  340 AL(1,J+1,K)=AL(1,J,K)-QT(1)/(SUM*(XL(1,J,K)+XL(1,J,K-1)))
  350 AL(1,1,K)=AL(1,N1,K)+QT(1)
      DO 360 J=1,N2
      AL(1,J,1)=AL(1,J,N4)-Q(1)
  360 AL(1,J,N5)=AL(1,J,2)+Q(1)
      DO 380 K=2,N4
      X1=HU/(ABS(SEAX)*SL2(2))
      SUMA(K)=SUMA(K)*X1
      SUMB(K)=SUMB(K)*X1
      SUMC(K)=SUMC(K)*X1
      EROAX=AMAX1(EROAX,ABS(SUMA(K)),ABS(SUMB(K)),ABS(SUMC(K)))
      IF (SEAX.GT.0.0) GO TO 370
      ROAX(K)=ROAX(K)+DT*SUMA(K)
      ROBX(K)=ROBX(K)+DT*SUMB(K)
      ROCX(K)=ROCX(K)+DT*SUMC(K)
      GO TO 380
  370 CONTINUE
      PAS1=DG1*(ROAX(K)*DH1+SUMA(K)-RA1*RPAX(K))
      PAS2=DG1*(ROBX(K)*DH1+SUMB(K)-RA1*RPBX(K))
      PAS3=DG1*(ROCX(K)*DH1+SUMC(K)-RA1*RPCX(K))
      RPAX(K)=ROAX(K)
      RPBX(K)=ROBX(K)
      RPCX(K)=ROCX(K)
      ROAX(K)=PAS1
      ROBX(K)=PAS2
      ROCX(K)=PAS3
  380 CONTINUE
      DO 390 K=2,N4
      SUMR(K)=SUMR(K)*HSU
      SUMZ(K)=SUMZ(K)*HSU
      X1=EL1*EH1(K)+EL2*EH2(K)
      X2=EM1*EJ1(K)+EM2*EJ2(K)
      AA=ABS((SUMR(K)-RQ*ANORM*X1)/SA3)
      BB=ABS((SUMZ(K)-RQ*ANORM*X2)/SA3)
      SUMR(K)=SUMR(K)/SA3
      SUMZ(K)=SUMZ(K)/SA3
      PRQ1=PRQ1+X1*(SUMR(K)-RA3*(RA(K)-RN(K)))+X2*(SUMZ(K)-RA3*(ZA(K)-ZN
     1 (K)))
      EAX=AMAX1(AA,BB)
      PAS1=DG3*(RA(K)*DH3-SUMR(K)-RA3*RN(K))
      PAS2=DG3*(ZA(K)*DH3-SUMZ(K)-RA3*ZN(K))
      RN(K)=RA(K)
      ZN(K)=ZA(K)
      RA(K)=PAS1
  390 ZA(K)=PAS2
      IF (NPLOT.GT.0) GO TO 560
      IF (IRQ.LT.0) GO TO 500
C         PROJECTION ONTO SUBSPACE FOR STABILITY ANALYSIS
```

```
        PNORM1=PNORM*DG1
        PRQ=(PRQ*HS*HU*HV+PRQ1*HV*ANORM)*PI
        PRQ=PRQ*DG1
        IF (NVAC.LT.0) GO TO 440
        PRQ2=0.0
        DO 400 J=2,N1
        DO 400 K=2,N4
  400   PRQ2=PRQ2+AKB(J,K)*EBD(J,K)
        PRQ=PRQ-HU*HV*PRQ2*BNORM*PI*DTB
        PNORM1=PNORM1+DTB*QNORM
        ERBO=0.0000001
        RQ1=(PRQ*BNORM)/PNORM1
        DO 420 J=2,N1
        DO 410 K=2,N4
        X1=AKB(J,K)+RQ1*EBD(J,K)
        AA=ABS(X1)
        ERBO=AMAX1(ERBO,AA)
  410   X(J,K)=X(J,K)+DTB*X1
        X(J,1)=X(J,N4)
  420   X(J,N5)=X(J,2)
        DO 430 K=1,N5
        X(1,K)=X(N1,K)
  430   X(N2,K)=X(2,K)
  440   CONTINUE
        RQ=PRQ/PNORM1
        DO 460 I=1,NI
        X3=RQ*SL2(I)*OP1(I)*OP2(I)/SA2
        DO 460 J=2,N1
        DO 450 K=2,N4
        X1=EAM4(I)*EG1(J,K)+EAM5(I)*EG2(J,K)+EAM6(I)*EG3(J,K)
        X1=-X1
  450   AL(I,J,K)=AL(I,J,K)+DG1*X1*X3
        AL(I,J,1)=AL(I,J,N4)-Q(I)
  460   AL(I,J,N5)=AL(I,J,2)+Q(I)
        DO 480 I=2,N3
        X3=RQ*RBO4*OP1(I)/SA1
        DO 480 J=2,N1
        DO 470 K=2,N4
        X1=EAM1(I)*EF1(J,K)+EAM2(I)*EF2(J,K)+EAM3(I)*EF3(J,K)
  470   RO(I,J,K)=RO(I,J,K)+DG1*X1*X3
        RO(I,J,1)=RO(I,J,N4)
  480   RO(I,J,N5)=RO(I,J,2)
        X3=RQ*ANORM/SA3
        DO 490 K=2,N4
        X1=EL1*EH1(K)+EL2*EH2(K)
        X2=EM1*EJ1(K)+EM2*EJ2(K)
        RA(K)=RA(K)+X3*X1*DG1
  490   ZA(K)=ZA(K)+X3*X2*DG1
  500   CONTINUE
C       PERIODICITY CONDITIONS
        DO 510 I=1,NI
        DO 510 K=1,N5
        AL(I,1,K)=AL(I,N1,K)+QT(I)
```

```
      XL(I,1,K)=XL(I,N1,K)+QT(I)
  510 AL(I,N2,K)=AL(I,2,K)-QT(I)
      DO 520 I=1,N3
      DO 520 K=1,N5
      RO(I,1,K)=RO(I,N1,K)
      XO(I,1,K)=XO(I,N1,K)
  520 RO(I,N2,K)=RO(I,2,K)
      RA(1)=RA(N4)
      RA(N5)=RA(2)
      ZA(1)=ZA(N4)
      X2=HU*HV*ZLE
C     PRESSURE COMPUTATION
      DO 530 I=1,N3
      PRP(I)=(AM(I)/(SPR(I)*X2))**GAM
  530 BMP(I)=PRP(I)*SPR(I)*HU*HV
      DO 540 I=2,NI
      PRM(I)=(AM(I)/(SPRM(I)*X2))**GAM
  540 BMM(I)=PRM(I)*SPRM(I)*HU*HV
      ZA(N5)=ZA(2)
      RN(1)=RA(1)
      ZN(1)=ZA(1)
C     FUNCTIONALS FOR COMPUTATION OF DESCENT COEFFICIENTS
      DO 550 K=2,N4
      X1=(RA(K)-RN(K))/DT
      X2=(ZA(K)-ZN(K))/DT
      AA1(3)=AA1(3)+X1*X1+X2*X2
      DO 550 J=2,N1
      DO 550 I=2,NI
      X1=(RO(I,J,K)-XO(I,J,K))/DT
      X2=(AL(I,J,K)-XL(I,J,K))/DT
      AA1(1)=AA1(1)+PQS(I)*X1*X1
  550 AA1(2)=AA1(2)+PQS(I)*X2*X2
      AA1(1)=AA1(1)*HS*HU*HV
      AA1(2)=AA1(2)*HS*HU*HV
      AA1(3)=AA1(3)*HV
  560 CONTINUE
      ENER=ENER*HS*HU*HV*ZLE
      RETURN
      END




      SUBROUTINE CIN (J,BJA,SJA)
C     EVALUATION OF THE COEFFICIENTS FOR THE PLASMA EQUATIONS SHOWING
C     HOW THEY DEPEND ON THE DIFFERENCE SCHEME
      USE NAME1
      USE NAME2
      USE NAME6
```

```
          USE NAME10
          USE NAME11
          USE NAME12
          USE NAME13
          ZZ=ZLE*ZLE
          ZL2=1.0/(2.0*ZZ)
          ZL4=1.0/(4.0*ZZ)
          GAM1=GAM-1.0
          HVV=2.0*HV*HV
          HUU=2.0*HU*HU
          HU2=2.0*HU
          HV2=2.0*HV
          DO 10 K=1,N4
          D1(K)=RB1(K)-0.5*(RA(K)+RA(K+1))
          D2(K)=ZB1(K)-0.5*(ZA(K)+ZA(K+1))
          D3(K)=(RA(K+1)-RA(K))/HV
       10 D4(K)=(ZA(K+1)-ZA(K))/HV
C         COMPUTE COEFFICIENTS FOR R COMING FROM BOUNDARY QUANTITIES
          DO 20 K=1,N4
          HB1(K)=D1(K)*ZBU1(K)-D2(K)*RBU1(K)
          CA1(K)=D1(K)*D1(K)+D2(K)*D2(K)
          CAF1(K)=D1(K)*D1(K)
          CAG1(K)=2.0*D1(K)*D2(K)
          CAH1(K)=D2(K)*D2(K)
          CB1(K)=RBU1(K)*RBU1(K)+ZBU1(K)*ZBU1(K)
          CC1(K)=D1(K)*RBU1(K)+D2(K)*ZBU1(K)
          CD1(K)=D3(K)*D3(K)+D4(K)*D4(K)
       20 CE1(K)=RBV1(K)*RBV1(K)+ZBV1(K)*ZBV1(K)
          DO 30 K=1,N4
          CF1(K)=D1(K)*D3(K)+D2(K)*D4(K)
          CG1(K)=D1(K)*RBV1(K)+D2(K)*ZBV1(K)
          CL1(K)=RBV1(K)*D3(K)+ZBV1(K)*D4(K)
          CM1(K)=RBU1(K)*RBV1(K)+ZBU1(K)*ZBV1(K)
       30 CN1(K)=D3(K)*RBU1(K)+D4(K)*ZBU1(K)
C         COMPUTE JACOBIAN
          DO 50 I=1,NI
          DO 40 K=2,N4
       40 D2(K)=RO(I,J,K)*RO(I,J,K)+RO(I,J+1,K)*RO(I,J+1,K)
          D2(N5)=D2(2)
          DO 50 K=2,N4
       50 E1(I,K)=D2(K)+D2(K+1)
          DO 60 I=1,N3
          X1=0.125*(1.0-PS(I))
          X2=0.125*PS(I)
          DO 60 K=2,N4
          D(I,K)=HB1(K)*(X1*E1(I,K)+X2*E1(I+1,K))
          BJA=AMIN1(BJA,D(I,K))
       60 SJA=AMAX1(SJA,D(I,K))
          DO 70 I=2,NI
          X1=0.125*(1.0+QS(I-1))
          X2=0.125*QS(I-1)
          DO 70 K=2,N4
          DM(I,K)=HB1(K)*(X1*E1(I,K)-X2*E1(I-1,K))
```

```
       BJA=AMIN1(BJA,DM(I,K))
    70 SJA=AMAX1(SJA,DM(I,K))
       IF (BJA.GT.0.0) GO TO 80
       PRINT 320
       STOP
C      PSI DERIVATIVES AND TOROIDAL FACTOR
    80 DO 90 I=1,NI
       X1=0.25*SL1(I)
       DO 90 K=2,N4
       V1(I,K)=((AL(I,J,K+1)-AL(I,J,K))*(AL(I,J,K+1)-AL(I,J,K))+(AL(I,J+1
      1 ,K+1)-AL(I,J+1,K))*(AL(I,J+1,K+1)-AL(I,J+1,K)))/HVV
       U1(I,K)=((AL(I,J+1,K)-AL(I,J,K))*(AL(I,J+1,K)-AL(I,J,K))+(AL(I,J+1
      1 ,K+1)-AL(I,J,K+1))*(AL(I,J+1,K+1)-AL(I,J,K+1)))/HUU
       UV1(I,K)=((AL(I,J+1,K+1)-AL(I,J,K))*(AL(I,J+1,K+1)-AL(I,J,K))-(AL
      1 (I,J+1,K)-AL(I,J,K+1))*(AL(I,J+1,K)-AL(I,J,K+1)))*HUV
       BK(I,K)=1.0+EP*(0.5*(RA(K)+RA(K+1))+X1*D1(K)*(RO(I,J,K)+RO(I,J+1,K
      1 )+RO(I,J,K+1)+RO(I,J+1,K+1)))
    90 CONTINUE
C      RECIPROCAL OF THE JACOBIAN
       DO 100 K=2,N4
       DD(1,K)=SA(1)/D(1,K)
   100 DD(NI,K)=SA(N3)/DM(NI,K)
       DO 110 I=2,N3
       DO 110 K=2,N4
   110 DD(I,K)=SA(I)/D(I,K)+SA(I-1)/DM(I,K)
       DO 120 K=2,N4
       XA1(K)=0.0
       XB1(K)=0.0
       XC1(K)=0.0
       XD1(K)=0.0
       XP1(K)=0.0
       XQ1(K)=0.0
       XG1(K)=0.0
       XN1(K)=0.0
   120 XF1(K)=0.0
       DO 210 I=1,NI
       X1=SL2(I)/HUU
       X2=0.25*SL2(I)
       X3=SL2(I)/HU2
       X4=SL2(I)/HV2
       X5=SL2(I)/HVV
       X7=0.25*SL1(I)
       X10=HUV*SL2(I)
       DO 130 K=2,N4
       D1(K)=X1*(RO(I,J+1,K)-RO(I,J,K))*(RO(I,J+1,K)-RO(I,J,K))
       D9(K)=RO(I,J,K)*RO(I,J,K)+RO(I,J+1,K)*RO(I,J+1,K)
       D3(K)=X3*(RO(I,J+1,K)*RO(I,J+1,K)-RO(I,J,K)*RO(I,J,K))
   130 D10(K)=(1.0-SL1(I)*RO(I,J,K))*(1.0-SL1(I)*RO(I,J,K))+(1.0-SL1(I)
      1 *RO(I,J+1,K))*(1.0-SL1(I)*RO(I,J+1,K))
       D1(N5)=D1(2)
       D9(N5)=D9(2)
       D3(N5)=D3(2)
       D10(N5)=D10(2)
```

```
      DO 140 K=2,N4
      D1(K)=D1(K)+D1(K+1)
      D2(K)=X2*(D9(K)+D9(K+1))
      D8(K)=X4*(D9(K+1)-D9(K))
      D3(K)=D3(K)+D3(K+1)
      D4(K)=0.25*(D10(K)+D10(K+1))
  140 D6(K)=(D10(K+1)-D10(K))/HV2
      DO 150 K=2,N4
      D5(K)=X5*((RO(I,J,K+1)-RO(I,J,K))*(RO(I,J,K+1)-RO(I,J,K))+(RO(I,J+
     1 1,K+1)-RO(I,J+1,K))*(RO(I,J+1,K+1)-RO(I,J+1,K)))
      D10(K)=((RO(I,J+1,K+1)-RO(I,J,K))*(RO(I,J+1,K+1)-RO(I,J,K))-(RO(I
     1 ,J+1,K)-RO(I,J,K+1))*(RO(I,J+1,K)-RO(I,J,K+1)))*X10
      D7(K)=X7*(RO(I,J,K)*(1.0-SL1(I)*RO(I,J,K))+RO(I,J+1,K)*(1.0-SL1(I)
     1 *RO(I,J+1,K)))
      D9(K)=(1.0-SL1(I)*RO(I,J+1,K))*(1.0-SL1(I)*RO(I,J+1,K))-(1.0-SL1(I
     1 )*RO(I,J,K))*(1.0-SL1(I)*RO(I,J,K))
  150 CONTINUE
      D7(N5)=D7(2)
      D9(N5)=D9(2)
      DO 160 K=2,N4
      D7(K)=D7(K)+D7(K+1)
      D9(K)=(D9(K)+D9(K+1))/HU2
      D11(K)=CA1(K)*D1(K)+CB1(K)*D2(K)+CC1(K)*D3(K)
      D12(K)=CD1(K)*D4(K)+CA1(K)*D5(K)+CE1(K)*D2(K)-CF1(K)*D6(K)+CG1(K)
     1 *D8(K)+2.0*D7(K)*CL1(K)+ZZ*BK(I,K)*BK(I,K)
      D13(K)=0.5*(-CF1(K)*D9(K)+CC1(K)*D8(K)+CG1(K)*D3(K))+CM1(K)*D2(K)
     1 +CA1(K)*D10(K)+CN1(K)*D7(K)
  160 CONTINUE
      DO 170 K=2,N4
  170 P1(I,K)=ZL2*(V1(I,K)*D11(K)+U1(I,K)*D12(K)-2.0*UV1(I,K)*D13(K))/BK
     1 (I,K)
      DO 180 K=2,N4
  180 Q1(I,K)=(P1(I,K)/BK(I,K)-U1(I,K))*DD(I,K)
C     COEFFICIENTS FOR THE PSI EQUATION
      DO 190 K=2,N4
      E1(I,K)=D11(K)*(DD(I,K)/BK(I,K))
      G1(I,K)=D12(K)*(DD(I,K)/BK(I,K))
      F1(I,K)=D13(K)*(DD(I,K)/BK(I,K))
      U1(I,K)=U1(I,K)*ZL2*(DD(I,K)/BK(I,K))
      V1(I,K)=V1(I,K)*ZL2*(DD(I,K)/BK(I,K))
  190 UV1(I,K)=UV1(I,K)*ZL2*(DD(I,K)/BK(I,K))
C     COEFFICIENTS FOR THE MAGNETIC AXIS EQUATION
      DO 200 K=2,N4
      XA1(K)=XA1(K)+0.5*(V1(I,K)*D1(K)+U1(I,K)*D5(K)-2.0*UV1(I,K)*D10(K)
     1 )
      XC1(K)=XC1(K)+0.5*(V1(I,K)*D3(K)-UV1(I,K)*D8(K))
      XG1(K)=XG1(K)+0.5*(U1(I,K)*D8(K)-UV1(I,K)*D3(K))
      XB1(K)=XB1(K)+U1(I,K)*D7(K)
      XN1(K)=XN1(K)-UV1(I,K)*D7(K)
      XD1(K)=XD1(K)+0.5*U1(I,K)*D4(K)
  200 XF1(K)=XF1(K)+0.5*(UV1(I,K)*D9(K)-U1(I,K)*D6(K))
  210 CONTINUE
      DO 240 I=1,N3
```

```
      X1=0.5*SA(I)
      X2=(BMP(I)+BMM(I+1))/GAM1
      DO 220 K=2,N4
  220 D1(K)=X1*(P1(I,K)/D(I,K)+P1(I+1,K)/DM(I+1,K)+X2)
C     ENERGY CALCULATION
      DO 230 K=2,N4
  230 ENER=ENER+D1(K)
  240 CONTINUE
      X1=SA(1)*PRP(1)
      X2=SA(N3)*PRM(NI)
C     COEFFICIENTS FOR THE R EQUATION
      DO 250 K=2,N4
      Q1(1,K)=Q1(1,K)+X1*D(1,K)
      P1(1,K)=P1(1,K)/(D(1,K)*D(1,K))+PRP(1)*BK(1,K)
      Q1(NI,K)=Q1(NI,K)+X2*DM(NI,K)
  250 PM1(NI,K)=P1(NI,K)/(DM(NI,K)*DM(NI,K))+PRM(NI)*BK(NI,K)
      DO 260 I=2,N3
      X1=SA(I)*PRP(I)
      X2=SA(I-1)*PRM(I)
      DO 260 K=2,N4
      Q1(I,K)=Q1(I,K)+X1*D(I,K)+X2*DM(I,K)
      PM1(I,K)=P1(I,K)/(DM(I,K)*DM(I,K))+PRM(I)*BK(I,K)
  260 P1(I,K)=P1(I,K)/(D(I,K)*D(I,K))+PRP(I)*BK(I,K)
      DO 270 I=1,N3
      D(I,1)=D(I,N4)
      X1=0.5*SA(I)
      DO 270 K=2,N4
      XP1(K)=XP1(K)+E4*Q1(I,K)*(1.0-0.25*SL1(I)*(RO(I,J,K)+RO(I,J+1,K)
     1 +RO(I,J,K+1)+RO(I,J+1,K+1)))
  270 XQ1(K)=XQ1(K)+X1*(P1(I,K)*D(I,K)+PM1(I+1,K)*DM(I+1,K))/HB1(K)
C     PERIODICITY CONDITIONS FOR THE COEFFICIENTS
      XA1(1)=XA1(N4)
      XB1(1)=XB1(N4)
      XC1(1)=XC1(N4)
      XG1(1)=XG1(N4)
      XD1(1)=XD1(N4)
      XF1(1)=XF1(N4)
      XN1(1)=XN1(N4)
      XP1(1)=XP1(N4)
      XQ1(1)=XQ1(N4)
      DO 280 I=1,NI
      U1(I,1)=U1(I,N4)
      V1(I,1)=V1(I,N4)
      UV1(I,1)=UV1(I,N4)
      E1(I,1)=E1(I,N4)
      F1(I,1)=F1(I,N4)
      G1(I,1)=G1(I,N4)
      P1(I,1)=P1(I,N4)
  280 Q1(I,1)=Q1(I,N4)
      DO 290 I=2,NI
      DM(I,1)=DM(I,N4)
  290 PM1(I,1)=PM1(I,N4)
      DO 310 K=2,N4
```

```
        DO 300 I=1,N3
  300 SPR(I)=SPR(I)+BK(I,K)*D(I,K)
        DO 310 I=2,NI
  310 SPRM(I)=SPRM(I)+BK(I,K)*DM(I,K)
        RETURN
C
  320 FORMAT (///,6X,17HNEGATIVE JACOBIAN)
        END




        SUBROUTINE CBO (J)
C       EVALUATES BOUNDARY QUANTITIES AT K+1/2
        USE NAME1
        USE NAME9
        USE NAME12
        USE NAME13
        IF (NVAC.LT.0) GO TO 30
C       EVALUATE R AND Z AND THEIR DERIVATIVES AT THE FREE BOUNDARY
        HU2=2.0*HU
        HV2=2.0*HV
        DO 10 K=2,N4
        D1(K)=0.25*(X(J,K)+X(J,K+1)+X(J+1,K)+X(J+1,K+1))
        D2(K)=(X(J+1,K)+X(J+1,K+1)-X(J,K)-X(J,K+1))/HU2
        D3(K)=(X(J,K+1)+X(J+1,K+1)-X(J,K)-X(J+1,K))/HV2
        D4(K)=0.5*(RVA(K)+RVA(K+1))
  10  D5(K)=0.5*(ZVA(K)+ZVA(K+1))
        DO 20 K=2,N4
        RB1(K)=D4(K)+D1(K)*(R(J,K)-D4(K))
        ZB1(K)=D5(K)+D1(K)*(Z(J,K)-D5(K))
        RBU1(K)=D2(K)*(R(J,K)-D4(K))+D1(K)*RU(J,K)
        ZBU1(K)=D2(K)*(Z(J,K)-D5(K))+D1(K)*ZU(J,K)
        RBV1(K)=(1.0-D1(K))*(RVA(K+1)-RVA(K))/HV+D3(K)*(R(J,K)-D4(K))+D1(K
     1 )*RV(J,K)
  20 ZBV1(K)=(1.0-D1(K))*(ZVA(K+1)-ZVA(K))/HV+D3(K)*(Z(J,K)-D5(K))+D1(K
     1 )*ZV(J,K)
        RB1(1)=RB1(N4)
        ZB1(1)=ZB1(N4)
        RBU1(1)=RBU1(N4)
        ZBU1(1)=ZBU1(N4)
        RBV1(1)=RBV1(N4)
        ZBV1(1)=ZBV1(N4)
        GO TO 50
C       PLASMA BOUNDARY CORRESPONDING TO OUTER WALL
  30 DO 40 K=1,N4
        RB1(K)=R(J,K)
        ZB1(K)=Z(J,K)
        RBU1(K)=RU(J,K)
```

```
          ZBU1(K)=ZU(J,K)
          RBV1(K)=RV(J,K)
   40 ZBV1(K)=ZV(J,K)
   50 CONTINUE
          RETURN
          END




          SUBROUTINE CDEN (CNOR,ICD)
C         COMPUTES PARALLEL CURRENT AND EVALUATES MERCIER CRITERION
          USE NAME1
          USE NAME2
          USE NAME6
          USE NAME7
          USE NAME8
          USE NAME10
          USE NAME12
          USE NAME13
          USE NAME14
          COMMON /CDEN/ DMIN(101), DMAX(101), RPMAX(101), RPMIN(101), CNORP(
        1 101), CNORT(101), BMAX(101), BMIN(101), ABB(101), AJJ(101), ABJ(1
        2 01), AZZ(101), ABZ(101), AW(101), AITP(101), BRIP(101), AMER(101)
        3 , AMERS(101), ASH(101), ASC(101), PLAM1(101), PLAM2(101), RTMAX(1
        4 01), RTMIN(101), TLAM(101), TLAM1(101), TLAM2(101), PLAM(101)
        5, FACP(101),ALP(101)
          USE NAME19
          USE NAME21
          USE NAME23
          REWIND 2
          XTOR=1.0E-12
          BJA=-1000.0
          SJA=1000.0
          ENERS=ENER
          DO 10 I=1,N3
          CNORP(I)=0.0
          CNORT(I)=0.0
   10 CNOR(I)=0.0
          DO 20 I=1,NI
          DMIN(I)=1000.0
          DMAX(I)=-1000.0
          RPMIN(I)=1000.0
          RPMAX(I)=-1000.0
          RTMIN(I)=1000.0
          BMAX(I)=-1000.0
          BMIN(I)=1000.0
          X1=-(PC(I)*QT(I)+TC(I)*Q(I))
          TLAM(I)=(TC(I)*(PC(I)-PC(1))+PC(I)*TC(I))/X1
```

```
 20 RTMAX(I)=-1000.0
    DO 30 I=2,N3
    ABB(I)=0.0
    AJJ(I)=0.0
    ABJ(I)=0.0
    AZZ(I)=0.0
    ABZ(I)=0.0
    AWW(I)=0.0
    Y3=2.0*PQS(I)*HS
    X1=QT(I)*QT(I)
    X2=PPRIM(I)
    X3=(QQ(I+1)-QQ(I-1))/Y3
    X4=QQ(I)
    FP(I)=X2
    X5=X1*X4*X4
    ASH(I)=X1*X3
    ASC(I)=X5*X5*X5
    AITP(I)=-(TC(I+1)-TC(I-1))/(QT(I)*Y3)
    FACP(I)=(TLAM(I+1)-TLAM(I-1))/Y3
    AW(I)=X1*X2*SPPVOL(I)
 30 CONTINUE
    DO 170 J=1,N1
    IF (J.EQ.1) GO TO 50
    DO 40 K=1,N4
    DO 40 I=1,NI
    DJ(I,K)=D(I,K)
    DJM(I,K)=DM(I,K)
    E2(I,K)=E1(I,K)
    G2(I,K)=G1(I,K)
    F2(I,K)=F1(I,K)
    P2(I,K)=P1(I,K)
    Q2(I,K)=Q1(I,K)
    V2(I,K)=V1(I,K)
    U2(I,K)=U1(I,K)
    GM(I,K)=GL(I,K)
    V4(I,K)=V3(I,K)
    V6(I,K)=V5(I,K)
    EL(I,K)=BK(I,K)
 40 UV2(I,K)=UV1(I,K)
    CALL CBO (J)
    GO TO 60
 50 CALL CBO (N1)
 60 CALL CIN (J,SJA,BJA)
    DO 70 K=1,N5
    D(NI,K)=1.0
 70 DM(1,K)=1.0
    DO 90 K=2,N4
    DMIN(1)=AMIN1(DMIN(1),D(1,K))
    DMAX(1)=AMAX1(DMAX(1),D(1,K))
    DMIN(NI)=AMIN1(DMIN(NI),DM(NI,K))
    DMAX(NI)=AMAX1(DMAX(NI),DM(NI,K))
C      COMPUTE B.B AND THE JACOBIAN RATIO
    DO 80 I=2,N3
```

```
      X1=(P1(I,K)/BK(I,K)-PRP(I))
      X2=(PM1(I,K)/BK(I,K)-PRM(I))
      X4=X1+X2
      GL(I,K)=ZLE*(BK(I,K)*D(I,K)*X1+BK(I,K)*DM(I,K)*X2)
      AWW(I)=AWW(I)+ZLE*HU*HV*(BK(I,K)*D(I,K)/X1+BK(I,K)*DM(I,K)/X2)*0.2
    1 5
      BMAX(I)=AMAX1(BMAX(I),X4)
      BMIN(I)=AMIN1(BMIN(I),X4)
      D1(I)=X4
      X3=0.5*(D(I,K)+DM(I,K))
      PHD(I,K)=BK(I,K)*ZLE*X3
      PHB2(I,K)=X4
      DMIN(I)=AMIN1(DMIN(I),X3)
   80 DMAX(I)=AMAX1(DMAX(I),X3)
      XO(NI,J,K)=D1(N3)
      DO 90 I=1,NI
   90 PM2(I,K)=0.25*(RO(I,J,K)+RO(I,J,K+1)+RO(I,J+1,K)+RO(I,J+1,K+1))
      DO 130 I=1,NI
      X9=2.0*ZLE
      X10=2.0*PQS(I)
      DO 100 K=2,N4
      P1(I,K)=E1(I,K)/DD(I,K)
      Q1(I,K)=F1(I,K)/DD(I,K)
  100 XUP(I,K)=G1(I,K)/DD(I,K)
      P1(I,1)=P1(I,N4)
      Q1(I,1)=Q1(I,N4)
      XUP(I,1)=XUP(I,N4)
      E1(I,1)=E1(I,N4)
      F1(I,1)=F1(I,N4)
      G1(I,1)=G1(I,N4)
      DO 120 K=2,N4
      PU=(AL(I,J+1,K+1)+AL(I,J+1,K)-AL(I,J,K)-AL(I,J,K+1))/(2.0*HU)
      PV=(AL(I,J+1,K+1)+AL(I,J,K+1)-AL(I,J,K)-AL(I,J+1,K))/(2.0*HV)
      X1=(AL(I,J,K+1)-AL(I,J,K))/HV
      X2=(AL(I,J+1,K)-AL(I,J,K))/HU
      X3=(AL(I,J+1,K+1)-AL(I,J+1,K))/HV
      X4=(AL(I,J+1,K+1)-AL(I,J,K+1))/HU
      Y4=(AL(I,J+1,K)+AL(I,J,K)-AL(I,J+1,K-1)-AL(I,J,K-1))/(2.0*HV)
      PHV1(I,K)=(F1(I,K)*PV+F1(I,K-1)*Y4-(G1(I,K)+G1(I,K-1))*X2)/
    1(4.0*PQS(I)*ZLE)
      V1(I,K)=(Q1(I,K)*PV/D(I,K)+Q1(I,K-1)*Y4/D(I,K-1)-(XUP(I,K)/D(I,K)
    1 +XUP(I,K-1)/D(I,K-1))*X2)/X9
      V5(I,K)=(Q1(I,K)*PV/DM(I,K)+Q1(I,K-1)*Y4/DM(I,K-1)-(XUP(I,K)/DM(I
    1 ,K)+XUP(I,K-1)/DM(I,K-1))*X2)/X9
      IF (J.EQ.1) GO TO 110
      Y2=(AL(I,J,K+1)+AL(I,J,K)-AL(I,J-1,K+1)-AL(I,J-1,K))/(2.0*HU)
      U1(I,K)=((P1(I,K)/D(I,K)+P2(I,K)/DJ(I,K))*X1-(Q1(I,K)*PU/D(I,K)+Q2
    1 (I,K)*Y2/DJ(I,K)))/X9
      PM1(I,K)=((P1(I,K)/DM(I,K)+P2(I,K)/DJM(I,K))*X1-(Q1(I,K)*PU/DM(I,K
    1 )+Q2(I,K)*Y2/DJM(I,K)))/X9
      PHU1(I,K)=((E1(I,K)+E2(I,K))*X1-F1(I,K)*PU-F2(I,K)*Y2)/
    1(4.0*PQS(I)*ZLE)
  110 CONTINUE
```

```
      YU=(RO(I,J+1,K)+RO(I,J+1,K+1)-RO(I,J,K)-RO(I,J,K+1))/(2.0*HU)
      YV=(RO(I,J,K+1)+RO(I,J+1,K+1)-RO(I,J,K)-RO(I,J+1,K))/(2.0*HV)
      X1=SL1(I)*(PM2(I,K)*CC1(K)+YU*CA1(K))
      X2=CF1(K)+SL1(I)*(YV*CA1(K)+PM2(I,K)*(CG1(K)-CF1(K)))
      Y1=2.0/(ZLE*PM2(I,K)*BK(I,K)*HB1(K))
      UV1(I,K)=Y1*(PV*X1-PU*X2)
      X3=(RA(K+1)-RA(K))/HV
      X4=(ZA(K+1)-ZA(K))/HV
      X5=RB1(K)-0.5*(RA(K)+RA(K+1))
      X6=ZB1(K)-0.5*(ZA(K)+ZA(K+1))
      Y3=YU*(X3*X6-X4*X5)+PM2(I,K)*(X3*ZBU1(K)-X4*RBU1(K))
      Y4=YV*(X5*ZBU1(K)-X6*RBU1(K))+YU*(RBV1(K)*X6-ZBV1(K)*X5)+PM2(I,K)*
     1 (RBV1(K)*ZBU1(K)-RBU1(K)*ZBV1(K))
      X12=0.5*(1.0/D(I,K)+1.0/DM(I,K))
      X7=SL1(I)*((1.0-SL1(I)*PM2(I,K))*Y3+SL1(I)*PM2(I,K)*Y4)*X12/(0
     1 .5*X9*BK(I,K))
      X8=P1(I,K)*BK(I,K)*X12*X12
      V3(I,K)=X7*X7+X8
  120 CONTINUE
  130 CONTINUE
      X1=0.25/ZLE
      DO 140 I=1,NI
      DJ(I,1)=DJ(I,N4)
      BK(I,1)=BK(I,N4)
      GL(I,1)=GL(I,N4)
      V3(I,1)=V3(I,N4)
      V5(I,1)=V5(I,N4)
      EL(I,1)=EL(I,N4)
      V1(I,1)=V1(I,N4)
      U1(I,1)=U1(I,N4)
      UV1(I,1)=UV1(I,N4)
      P1(I,1)=P1(I,N4)
      Q1(I,1)=Q1(I,N4)
      PM1(I,1)=PM1(I,N4)
  140 CONTINUE
      IF (J.EQ.1) GO TO 170
      DO 160 I=2,N3
      X9=SL1(I)*2.0
      X10=2.0*ZLE*PQS(I)
      Y9=2.0*SL2(I)-SL2(I+1)
      Y10=2.0*SL2(I)-SL2(I-1)
      Y11=0.25*(SL2(I+1)+2.0*SL2(I)+SL2(I-1))
      X12=2.0*PQS(I)*HS*Y11
      DO 150 K=2,N4
      X1=(UV1(I,K)+UV1(I,K-1)-UV2(I,K)-UV2(I,K-1))/(2.0*HU*X9)
      X2=(UV1(I,K)+UV2(I,K)-UV1(I,K-1)-UV2(I,K-1))/(2.0*HV*X9)
      Y1=0.5*(V5(I+1,K)+V6(I+1,K))
      Y2=0.5*(V1(I,K)+V2(I,K))
      Y3=0.5*(V5(I,K)+V6(I,K))
      Y4=0.5*(V1(I-1,K)+V2(I-1,K))
      X3=0.25*(Y1+Y2+Y3+Y4)
      X5=((Y1+Y2)-(Y3+Y4))/(SL2(I+1)-SL2(I-1))
      Y1=0.5*(PM1(I+1,K)+PM1(I+1,K-1))
```

```
          Y2=0.5*(U1(I,K)+U1(I,K-1))
          Y3=0.5*(PM1(I,K)+PM1(I,K-1))
          Y4=0.5*(U1(I-1,K)+U1(I-1,K-1))
          X4=0.25*(Y1+Y2+Y3+Y4)
          X6=((Y1+Y2)-(Y3+Y4))/(SL2(I+1)-SL2(I-1))
          X7=0.25*(GL(I,K)+GL(I,K-1)+GM(I,K)+GM(I,K-1))
C         COMPUTE PARALLEL CURRENT
          CPOL=-(X3*(X1-X6)+X4*(X5-X2))/X7
          X8=0.25*(V3(I,K)+V3(I,K-1)+V4(I,K)+V4(I,K-1))
C         COMPUTE INTEGRALS FOR MERCIER CRITERION
          ABB(I)=ABB(I)+HU*HV*X7/X8
          ABJ(I)=ABJ(I)+HU*HV*CPOL*X7/X8
          AZZ(I)=AZZ(I)+HU*HV*CPOL*CPOL*X7/X8
          XL(I,J,K)=CPOL
  150 CONTINUE
  160 CONTINUE
          WRITE(2) ((PHV1(I,K),I=2,N3),K=2,N4),((PHU1(I,K),I=2,N3),K=2,N4),
         1((PHD(I,K),I=2,N3),K=2,N4),((PHB2(I,K),I=2,N3),K=2,N4),
         1((GL(I,K),I=2,N3),K=2,N4),((V3(I,K),I=2,N3),K=2,N4)
  170 CONTINUE
C         COMPUTE MIRROR RATIO
          DO 180 I=2,N3
          BMIN(I)=SQRT(BMIN(I))
          BMAX(I)=SQRT(BMAX(I))
          BS4(I)=AWW(I)
  180 BRIP(I)=(BMAX(I)-BMIN(I))/(BMAX(I)+BMIN(I))
          DO 210 I=2,N3
          TLAM(I)=0.0
          DO 200 J=2,N1
          DO 200 K=2,N4
  200 TLAM(I)=TLAM(I)+XL(I,J,K)
          TLAM(I)=TLAM(I)*HU*HV
  210 SUMB(I)=0.0
          CALL FPHI(ABBF,ABJF,AZZF,ALP,FP,FACP)
          DO 211 I=2,N3
          SUMB(I)=0.0
          DO 211 J=2,N1
          DO 211 K=2,N4
  211 SUMB(I)=SUMB(I)+XO(I,J,K)*HU*HV
          DO 212 I=2,N3
          Y2=FP(I)
          ABZ(I)=ABJ(I)-AITP(I)*ABB(I)
          ABZN(I)=ABJF(I)-AITP(I)*ABBF(I)
          BS2(I)=(BS4(I)-BS2(I))/BS4(I)
          IF (ABS(QQ(I)).LT.0.000001) QQ(I)=0.000001
          X1=SNL2(I)*(RBOU**4.0)/(QQ(I)*QQ(I))
          X2=QT(I)*QT(I)*QT(I)
          X1=X1/(X2*X2)
          X1=ABS(X1)
          X7=X1/SL2(I)
          ASC(I)=ASC(I)/X1
          ABZ(I)=-ASH(I)*ABZ(I)*X1
          ABZN(I)=-ASH(I)*ABZN(I)*X1
```

```
      ASH(I)=0.25*ASH(I)*ASH(I)*X1
      PAS=AW(I)
      AW(I)=(AW(I)-Y2*Y2*AWW(I))*ABB(I)*X1
      AWN(I)=(PAS-Y2*Y2*AWW(I))*ABBF(I)*X1
      ABJ(I)=X1*(ABJ(I)*ABJ(I)-ABB(I)*AZZ(I))
      ABJN(I)=X1*(ABJF(I)*ABJF(I)-ABBF(I)*AZZF(I))
      AMERS(I)=ASH(I)+ABZ(I)+AW(I)
      AMER(I)=AMERS(I)+ABJ(I)
      AMERN(I)=ABJN(I)+ASH(I)+ABZN(I)+AWN(I)
  212 CONTINUE
      ENER=ENERS
C     COMPUTE NORM AND FOURIER COEFFICIENTS
      DO 230 I=2,N3
      DO 220 J=2,N1
      DO 220 K=2,N4
      X1=XL(I,J,K)-TLAM(I)
      X2=(XO(I,J,K)-SUMB(I))*FP(I)
      CNOR(I)=CNOR(I)+X2*X2
      RTMIN(I)=AMIN1(RTMIN(I),X1)
      RTMAX(I)=AMAX1(RTMAX(I),X1)
  220 CNORT(I)=CNORT(I)+X1*X1
      X1=CNOR(I)*HU*HV
      X3=CNORT(I)*HU*HV
      CNOR(I)=SQRT(X1)/4.0
  230 CNORT(I)=SQRT(X3)/4.0
      DO 240 J=1,N2
      U=2.0*PI*(J-2.5)*HU
      UY=2.0*PI*(J-2.0)*HU
      YUP(1,J)=1.0
      YUP(2,J)=SIN(MIS*UY)
      YUP(3,J)=COS(MIS*UY)
      XUP(1,J)=1.0
      XUP(2,J)=SIN(MIS*U)
  240 XUP(3,J)=COS(MIS*U)
      DO 250 K=1,N5
      V=2.0*PI*(K-2.5)*HV
      VY=2.0*PI*(K-2.0)*HV
      YVP(1,K)=1.0
      YVP(2,K)=SIN(NIS*VY)
      YVP(3,K)=COS(NIS*VY)
      XVP(1,K)=1.0
      XVP(2,K)=SIN(NIS*V)
  250 XVP(3,K)=COS(NIS*V)
      DO 320 I=2,N3
      DO 280 K=2,N4
      DO 270 L=1,3
      SFI(L,K)=0.0
      SFI(L+3,K)=0.0
      DO 260 J=2,N1
      SFI(L,K)=SFI(L,K)+XL(I,J,K)*XUP(L,J)
  260 SFI(L+3,K)=SFI(L+3,K)+XO(I,J,K)*YUP(L,J)*FP(I)
      SFI(L,K)=SFI(L,K)*2.0*HU
  270 SFI(L+3,K)=SFI(L+3,K)*2.0*HU
```

```
      SFI(1,K)=0.5*SFI(1,K)
  280 SFI(4,K)=0.5*SFI(4,K)
      DO 310 L=1,3
      DO 300 M=1,3
      SRO(L,M)=0.0
      SRO(L+3,M)=0.0
      DO 290 K=2,N4
      SRO(L+3,M)=SRO(L+3,M)+SFI(L+3,K)*YVP(M,K)
  290 SRO(L,M)=SRO(L,M)+SFI(L,K)*XVP(M,K)
      SRO(L+3,M)=SRO(L+3,M)*2.0*HV
  300 SRO(L,M)=SRO(L,M)*2.0*HV
      SRO(L+3,1)=SRO(L+3,1)*0.5
  310 SRO(L,1)=SRO(L,1)*0.5
      TLAM1(I)=0.5*(SRO(3,3)+SRO(2,2))
      TLAM2(I)=0.5*(SRO(2,3)-SRO(3,2))
      PLAM(I)=FP(I)*SUMB(I)+FACP(I)
      PLAM1(I)=0.5*(SRO(6,3)+SRO(5,2))
      PLAM2(I)=0.5*(SRO(5,3)-SRO(6,2))
  320 CONTINUE
      RETURN
      END




      SUBROUTINE FPHI(ABBF,ABJF,AZZF,ALP,FP,FACP)
C     COMPUTES FOURIER COEFFICIENTS OF FIELD AND CURRENT IN FLUX
C     COORDINATES FOR MERCIER CRITERION AND TRANSPORT
      USE NAME21
      USE NAME22
      USE NAME12
      USE NAME10
      USE NAME1
      DIMENSION FP(101),ABBF(101),ABJF(101),AZZF(101)
      DIMENSION SCL(4),ALP(101),FACP(101)
      ERPC=0.0
      ERTC=0.0
      MFT=MF
      NFT=NF
      FAC=2.0*PI/(FLOAT(NJ))
      FILJ(1)=1.0
      FILK(1)=1.0
      DO 10 J=2,MFT
   10 FILJ(J)=SIN((J-1)*FAC)/(FAC*(J-1))
      FAC=2.0*PI/FLOAT(NK)
      DO 20 K=2,NFT
   20 FILK(K)=SIN((K-1)*FAC)/(FAC*(K-1))
      DO 300 I=2,N3
      FAC=-4.0/(QT(I)*PC(I)+Q(I)*TC(I))
```

```
      BS4(I)=0.0
      BSAV(I)=0.0
      REWIND 2
      SAV=0.0
      DO 30 J=2,N1
      READ(2) ((PHV1(I9,K),I9=2,N3),K=2,N4),((PHU1(I9,K),I9=2,N3),K=2,
     1N4),((PHD(I9,K),I9=2,N3),K=2,N4),((PHB2(I9,K),I9=2,N3),K=2,N4),
     1((PGL(I9,K),I9=2,N3),K=2,N4),((PV3(I9,K),I9=2,N3),K=2,N4)
      DO 30 K=2,N4
      U1(J,K)=PHU1(I,K)
      V1(J,K)=PHV1(I,K)
      D(J,K)=PHD(I,K)
      SAV=SAV+HU*HV*PHD(I,K)
      B2(J,K)=PHB2(I,K)
      GL(J,K)=PGL(I,K)
      BS4(I)=BS4(I)+0.25*FAC*HU*HV*D(J,K)/B2(J,K)
      BSAV(I)=BSAV(I)+B2(J,K)*HU*HV
   30 V3(J,K)=PV3(I,K)
      BSAV(I)=SQRT(BSAV(I))
      SAV1=SAV*FAC
C     COMPUTE SCALAR POTENTIAL PHI
      PHI(1,1)=0.0
      DO 40 K=2,N4
   40 PHI(1,K)=PHI(1,K-1)-HV*V1(N1,K)
      DO 60 J=2,N1
      U1(J,1)=U1(J,N4)
      DO 50 K=1,N4
   50 PHI(J,K)=PHI(J-1,K)-HU*U1(J,K)
   60 PHPC(J)=PHI(J,N4)-PHI(J,1)
      DO 70 K=2,N4
   70 PHTC(K)=PHI(N1,K)-PHI(1,K)
      SUMPC=0.0
      SUMTC=0.0
      DO 80 J=2,N1
      AA=ABS(PHPC(J)-PC(I))
      ERPC=AMAX1(ERPC,AA)
   80 SUMPC=SUMPC+HU*PHPC(J)
      DO 90 K=2,N4
      AA=ABS(PHTC(K)-TC(I))
      ERTC=AMAX1(ERTC,AA)
   90 SUMTC=SUMTC+HV*PHTC(K)
      PC(I)=SUMPC
      TC(I)=SUMTC
C     NORMALIZE PERIODS OF POTENTIAL PHI AND FLUX PSI
      DO 100 J=2,N1
      DO 100 K=2,N4
      X1=PHI(J,K)
      X2=0.25*(AL(I,J,K)+AL(I,J+1,K)+AL(I,J,K+1)+AL(I,J+1,K+1))
      X3=PC(I)+Q(I)*TC(I)/QT(I)
      PHI(J,K)=2.0*PI*(X1+TC(I)*X2/QT(I))/X3
  100 PSI(J,K)=-2.0*PI*(X2-Q(I)*X1/PC(I))*PC(I)/(QT(I)*X3)
      SUM1=0.0
      SUM2=0.0
```

```
      DO 110 J=2,N1
  110 SUM1=SUM1+HU*PHI(J,2)
      DO 120 K=2,N4
  120 SUM2=SUM2+HV*PSI(2,K)
      DO 130 J=2,N1
      DO 130 K=2,N4
      PHI(J,K)=PHI(J,K)-SUM1
  130 PSI(J,K)=PSI(J,K)-SUM2
      DO 140 J=2,N1
  140 XL(I,J,N5)=XL(I,J,2)
      DO 150 K=2,N5
  150 XL(I,N2,K)=XL(I,2,K)
      FAC1=FAC*HU*HV
      DO 301 NM=1,1
      TRAP(I,NM)=0.0
      SNLP(I,NM)=0.0
      X1=0.1+(NM-1)*0.2
      BMAGM(NM)=1.0-X1*X1
      IF(NM.EQ.1) BMAGM(NM)=AMAGM
      DO 190 M=1,MFT
      DO 190 N=1,NFT
      DO 160 L=1,4
      SCL(L)=0.0
      BB(M,N,L)=0.0
      OL(M,N,L)=0.0
  160 BL(M,N,L)=0.0
C     COMPUTE FOURIER COEFFICIENTS OF 1/(B.B) AND PARALLEL CURRENT
      DO 180 J=2,N1
      DO 180 K=2,N4
      X1=COS((M-1)*PSI(J,K))
      X2=SIN((M-1)*PSI(J,K))
      X3=COS((N-1)*PHI(J,K))
      X4=SIN((N-1)*PHI(J,K))
      CL(1)=X1*X3
      CL(2)=X1*X4
      CL(3)=X2*X3
      CL(4)=X2*X4
      X5=SQRT(B2(J,K))
      X6=BMAGM(NM)*X5/BSAV(I)
      FAC3=SQRT(AMAX1(0.0,1.0-X6))/X5
      DO 170 L=1,4
      FAC2=FAC1*D(J,K)*CL(L)
      SCL(L)=SCL(L)+FAC2*B2(J,K)
      BB(M,N,L)=BB(M,N,L)+FAC2
  170 OL(M,N,L)=OL(M,N,L)+FAC2*B2(J,K)*FAC3
  180 CONTINUE
  190 CONTINUE
      DO 191 L=1,4
      BL(1,1,L)=0.0
  191 RL(1,1,L)=0.0
      DO 200 M=1,MFT
      DO 200 L=1,4
      BB(M,1,L)=BB(M,1,L)/2.0
```

```
  200 OL(M,1,L)=OL(M,1,L)/2.0
      DO 210 N=1,NFT
      DO 210 L=1,4
      BB(1,N,L)=BB(1,N,L)/2.0
  210 OL(1,N,L)=OL(1,N,L)/2.0
      BS2(I)=0.0
      DO 240 L=1,4
      BS2(I)=BS2(I)+BB(1,1,L)*BB(1,1,L)
      DO 220 M=2,MFT
  220 BS2(I)=BS2(I)+0.5*BB(M,1,L)*BB(M,1,L)
      DO 230 N=2,NFT
  230 BS2(I)=BS2(I)+0.5*BB(1,N,L)*BB(1,N,L)
      DO 240 M=2,MFT
      DO 240 N=2,NFT
  240  BS2(I)=BS2(I)+0.25*BB(M,N,L)*BB(M,N,L)
      AMUL=0.000001
C     COMPUTE FOURIER COEFFICIENTS OF PARALLEL CURREMT FROM
C     FOURIER COEFFICIENTS OF 1/(B.B)
      DO 260 M=1,MFT
      DO 260 N=1,NFT
      X1=-((M-1)*PC(I)+(N-1)*TC(I))/QT(I)
      X3=-((M-1)*PC(I)-(N-1)*TC(I))/QT(I)
      X4=(N-1.0)+(M-1.0)*Q(I)/QT(I)
      IF(ABS(X4).LT.AMUL)X4=SIGN(AMUL,X4)
      X2=(N-1.)-(M-1.)*Q(I)/QT(I)
      IF(ABS(X2).LT.AMUL)X2=SIGN(AMUL,X2)
      IF(M.EQ.1.AND.N.EQ.1) GO TO 250
      Y1=0.5*(BB(M,N,1)+BB(M,N,4))
      Y2=0.5*(BB(M,N,1)-BB(M,N,4))
      Y3=0.5*(BB(M,N,3)+BB(M,N,2))
      Y4=0.5*(BB(M,N,3)-BB(M,N,2))
      BL(M,N,1)=X1*Y1/X2-X3*Y2/X4
      BL(M,N,4)=X1*Y1/X2+X3*Y2/X4
      BL(M,N,3)=-X3*Y3/X4+X1*Y4/X2
      BL(M,N,2)=-X3*Y3/X4-X1*Y4/X2
      Y1=0.5*(OL(M,N,1)+OL(M,N,4))
      Y2=0.5*(OL(M,N,1)-OL(M,N,4))
      Y3=0.5*(OL(M,N,3)+OL(M,N,2))
      Y4=0.5*(OL(M,N,3)-OL(M,N,2))
      RL(M,N,1)=(X1*Y1/X2-X3*Y2/X4)
      RL(M,N,4)=(X1*Y1/X2+X3*Y2/X4)
      RL(M,N,3)=(-X3*Y3/X4+X1*Y4/X2)
      RL(M,N,2)=(-X3*Y3/X4-X1*Y4/X2)
  250 CONTINUE
      ABI(M,N,I)=BB(M,N,1)
      BBI(M,N,I)=BB(M,N,2)
      CBI(M,N,I)=BB(M,N,3)
      DBI(M,N,I)=BB(M,N,4)
      AOL(M,N,I)=OL(M,N,1)
      BOL(M,N,I)=OL(M,N,2)
      COL(M,N,I)=OL(M,N,3)
      DOL(M,N,I)=OL(M,N,4)
      ABL(M,N,I)=BL(M,N,1)
```

```
          BBL(M,N,I)=BL(M,N,2)
          CBL(M,N,I)=BL(M,N,3)
          DBL(M,N,I)=BL(M,N,4)
          ARL(M,N,I)=RL(M,N,1)
          BRL(M,N,I)=RL(M,N,2)
          CRL(M,N,I)=RL(M,N,3)
          DRL(M,N,I)=RL(M,N,4)
      260 CONTINUE
          ABBF(I)=0.
          ABJF(I)=0.0
          AZZF(I)=0.0
          ALP(I)=0.0
C         COMPUTE PARALLEL CURRENT AND GUIDING CENTER STEP SIZE
          DO 290 J=2,N1
          DO 290 K=2,N4
          U1(J,K)=0.0
          XO(I,J,K)=0.0
          DO 280 M=1,MFT
          DO 280 N=1,NFT
          X1=COS((M-1)*PSI(J,K))
          X2=SIN((M-1)*PSI(J,K))
          X3=COS((N-1)*PHI(J,K))
          X4=SIN((N-1)*PHI(J,K))
          CL(1)=X1*X3
          CL(2)=X1*X4
          CL(3)=X2*X3
          CL(4)=X2*X4
          DO 270 L=1,4
          U1(J,K)=U1(J,K)+CL(L)*RL(M,N,L)
      270 XO(I,J,K)=XO(I,J,K)+CL(L)*BL(M,N,L)
      280 CONTINUE
          X1=HU*HV*GL(J,K)/V3(J,K)
          X2=XO(I,J,K)*FP(I)+FACP(I)
C         COMPUTE INTEGRALS FOR THE MERCIER CRITERION
          ABBF(I)=ABBF(I)+X1
          ABJF(I)=ABJF(I)+X1*X2
          AZZF(I)=AZZF(I)+X1*X2*X2
          ALP(I)=ALP(I)+X1*U1(J,K)*U1(J,K)/PC(I)
          X5=SQRT(B2(J,K))
          X6=BMAGM(NM)*X5/BSAV(I)
          IF(1.0-X6.LT.0.0)TRAP(I,NM)=TRAP(I,NM)+HU*HV
      290 CONTINUE
          SNLP(I,NM)=ALP(I)
      301 CONTINUE
      300 CONTINUE
          RETURN
          END
```

```
      SUBROUTINE TSOR (ERVA,EVNE,IT,ITER)
C     COMPUTES ITERATION OF 3D VACUUM EQUATIONS
      USE NAME1
      USE NAME4
      USE NAME9
      USE NAME11
      USE NAME12
      USE NAME13
      NCO=0
      UBI=1.0-USM
      PM1=1.0-OM1
      PM2=1.0-OM2
      AT=0.1
      IF (NPLOT.LT.0) GO TO 20
      DO 10 I=1,NIV
      BPV(I)=0.0
   10 BTV(I)=0.0
      GO TO 40
   20 CONTINUE
   30 CONTINUE
      NCO=NCO+1
      A11=0.0
      A22=0.0
      A12=0.0
      AA1(4)=0.0
      AA1(5)=0.0
      ERVA=0.0
   40 CONTINUE
      ZZ=ZLE*ZLE
      DU=1.0/(4.0*HU)
      DV=1.0/(4.0*HV)
      DSU=1.0/(2.0*HR*HU)
      DSV=1.0/(2.0*HR*HV)
      DVU=1.0/(2.0*HV*HU)
      DUV=DVU
      H4=1.0/(HV*HV)
      HH1=0.5*H1
      HH2=0.5*H2
      HH4=0.5*H4
      CALL CBO (N1)
      CALL CV (1)
C     COEFFICIENT MATRICES HAVE THE INDEX 1 WHEN EVALUATED AT K+1/2
C     AND THE INDEX 2 WHEN EVALUATED AT K-1/2
      DO 310 J=2,N1
      DO 50 K=1,N4
      RB2(K)=RB1(K)
      ZB2(K)=ZB1(K)
      RBU2(K)=RBU1(K)
      ZBU2(K)=ZBU1(K)
      RBV2(K)=RBV1(K)
   50 ZBV2(K)=ZBV1(K)
      DO 60 I=1,N6
      DO 60 K=1,N4
```

```
            VA2(I,K)=VA1(I,K)
            VB2(I,K)=VB1(I,K)
            VC2(I,K)=VC1(I,K)
            VE2(I,K)=VE1(I,K)
            VL2(I,K)=VL1(I,K)
     60     VD2(I,K)=VD1(I,K)
            CALL CBO (J)
            CALL CV (J)
            IF (NPLOT.LT.0) GO TO 80
C           COMPUTE FINAL VALUES OF VACUUM MAGNETIC FIELD
            DO 70 K=2,N4
            DO 70 I=1,N6
            S=(I-0.5)*HR
            S1=1.0-S
            X1=S1*RBU1(K)+S*RU(J,K)
            X2=S1*RBV1(K)+S*RV(J,K)
            X3=S1*ZBU1(K)+S*ZU(J,K)
            X4=S1*ZBV1(K)+S*ZV(J,K)
            AK=1.0+EP*(RB1(K)+S*(R(J,K)-RB1(K)))
            TU=DU*(PT(I,J+1,K)+PT(I,J+1,K+1)+PT(I+1,J+1,K)+PT(I+1,J+1,K+1)-PT
           1 (I,J,K)-PT(I+1,J,K)-PT(I,J,K+1)-PT(I+1,J,K+1))
            TV=DV*(PT(I,J,K+1)+PT(I+1,J,K+1)+PT(I,J+1,K+1)+PT(I+1,J+1,K+1)-PT
           1 (I,J,K)-PT(I+1,J,K)-PT(I,J+1,K)-PT(I+1,J+1,K))
            PU=TU
            PV=TV
            X1=SQRT(X1*X1+X3*X3)
            X2=SQRT(X2*X2+X4*X4+ZZ*AK*AK)
C           POLOIDAL AND TOROIDAL VACUUM FIELDS
            BPV(I)=BPV(I)+PU/X1
     70     BTV(I)=BTV(I)+PV/X2
            GO TO 310
     80     CONTINUE
C           SUCCESSIVE OVERRELAXATION AND ITERATION FOR
C           FREE BOUNDARY POINTS
            DO 90 K=2,N4
            D2(K)=VA1(1,K)+VA1(1,K-1)+VA2(1,K)+VA2(1,K-1)
            D3(K)=VB1(1,K)+VB2(1,K)
            D4(K)=VC1(1,K)+VC1(1,K-1)
            D5(K)=VC2(1,K)+VC2(1,K-1)
            D6(K)=VD1(1,K)+VD1(1,K-1)
            D7(K)=VD2(1,K)+VD2(1,K-1)
     90     D8(K)=VE1(1,K)+VE2(1,K)
            D8(1)=D8(N4)
            D3(1)=D3(N4)
            DO 100 K=2,N4
    100     D1(K)=HH1*D2(K)+HH4*(D3(K)+D3(K-1))+HH2*(D4(K)+D5(K))+DSU*(D6(K)
           1 -D7(K))+DSV*(D8(K)-D8(K-1))+DVU*(VL1(1,K)-VL1(1,K-1)-VL2(1,K)+VL2
           2 (1,K-1))
            DO 110 K=2,N4
    110     D10(K)=HH1*D2(K)*PT(2,J,K)+HH2*(D4(K)*PT(1,J+1,K)+D5(K)*PT(1,J-1,K
           1 ))+DSU*(D6(K)*PT(2,J+1,K)-D7(K)*PT(2,J-1,K))+DSV*(D8(K)*PT(2,J,K+
           2 1)-D8(K-1)*PT(2,J,K-1))+DVU*(VL1(1,K)*PT(1,J+1,K+1)-VL1(1,K-1)*PT
           3 (1,J+1,K-1)-VL2(1,K)*PT(1,J-1,K+1)+VL2(1,K-1)*PT(1,J-1,K-1))
```

```
      DO 120 K=2,N4
      SUM2=D10(K)+HH4*(D3(K)*PT(1,J,K+1)+D3(K-1)*PT(1,J,K-1))
      PAS2=PM2*PT(1,J,K)+(OM2*SUM2)/D1(K)
      BB=ABS(PAS2-PT(1,J,K))/DT
      AA1(5)=AA1(5)+BB*BB
      ERVA=AMAX1(ERVA,BB)
      PT(1,J,K)=PAS2
  120 CONTINUE
C     PERIODICITY CONDITIONS
      PT(1,J,1)=PT(1,J,N4)-C2
      PT(1,J,N5)=PT(1,J,2)+C2
C     ITERATION FOR INTERIOR POINTS
      DO 190 I=2,N6
      DO 130 K=2,N4
      D12(K)=D2(K)
      D3(K)=D6(K)
      D4(K)=D7(K)
      D5(K)=D8(K)
  130 CONTINUE
      DO 140 K=2,N4
      D2(K)=VA1(I,K)+VA1(I,K-1)+VA2(I,K)+VA2(I,K-1)
      D6(K)=VD1(I,K)+VD1(I,K-1)
      D7(K)=VD2(I,K)+VD2(I,K-1)
      D8(K)=VE1(I,K)+VE2(I,K)
      D9(K)=VB1(I,K)+VB2(I,K)+VB1(I-1,K)+VB2(I-1,K)
  140 CONTINUE
      DO 150 K=2,N4
      D10(K)=VC1(I,K)+VC1(I,K-1)+VC1(I-1,K)+VC1(I-1,K-1)
      D11(K)=VC2(I,K)+VC2(I,K-1)+VC2(I-1,K)+VC2(I-1,K-1)
      D13(K)=VL1(I,K)+VL1(I-1,K)
  150 D14(K)=VL2(I,K)+VL2(I-1,K)
      D5(1)=D5(N4)
      D13(1)=D13(N4)
      D14(1)=D14(N4)
      D8(1)=D8(N4)
      D9(1)=D9(N4)
      DO 160 K=2,N4
      D1(K)=HH1*(D2(K)+D12(K))+HH4*(D9(K)+D9(K-1))+HH2*(D10(K)+D11(K))
     1 +DSU*(D6(K)-D7(K)-D3(K)+D4(K))+DSV*(D8(K)-D8(K-1)-D5(K)+D5(K-1))
     2 +DUV*(D13(K)-D13(K-1)-D14(K)+D14(K-1))
  160 CONTINUE
      DO 170 K=2,N4
  170 D15(K)=HH1*(D2(K)*PT(I+1,J,K)+D12(K)*PT(I-1,J,K))+HH2*(D10(K)*PT(I
     1 ,J+1,K)+D11(K)*PT(I,J-1,K))+DSU*(D6(K)*PT(I+1,J+1,K)-D7(K)*PT(I+1
     2 ,J-1,K)-D3(K)*PT(I-1,J+1,K)+D4(K)*PT(I-1,J-1,K))+DSV*(D8(K)*PT(I+
     3 1,J,K+1)-D8(K-1)*PT(I+1,J,K-1)-D5(K)*PT(I-1,J,K+1)+D5(K-1)*PT(I-1
     4 ,J,K-1))+DUV*(D13(K)*PT(I,J+1,K+1)-D13(K-1)*PT(I,J+1,K-1)-D14(K)
     5 *PT(I,J-1,K+1)+D14(K-1)*PT(I,J-1,K-1))
      DO 180 K=2,N4
      SUM2=D15(K)+HH4*(D9(K)*PT(I,J,K+1)+D9(K-1)*PT(I,J,K-1))
      PAS2=PM2*PT(I,J,K)+(OM2*SUM2)/D1(K)
      BB=ABS(PAS2-PT(I,J,K))/DT
      AA1(5)=AA1(5)+BB*BB
```

```
        ERVA=AMAX1(ERVA,BB)
        PT(I,J,K)=PAS2
  180 CONTINUE
C       PERIODICITY CONDITIONS
        PT(I,J,1)=PT(I,J,N4)-C2
  190 PT(I,J,N5)=PT(I,J,2)+C2
        U=(J-2.5)*HU
        IF (U.GT.USM.AND.U.LT.UBI) GO TO 230
        DO 200 K=2,N4
        D9(K)=VB1(N6,K)+VB2(N6,K)
        D10(K)=VC1(N6,K)+VC1(N6,K-1)
  200 D11(K)=VC2(N6,K)+VC2(N6,K-1)
        D9(1)=D9(N4)
        DO 210 K=2,N4
        D1(K)=HH1*D2(K)+HH4*(D9(K)+D9(K-1))+HH2*(D10(K)+D11(K))+DSU*(-D6(K
     1 )+D7(K))+DSV*(-D8(K)+D8(K-1))+DUV*(VL1(N6,K)-VL1(N6,K-1)-VL2(N6,K
     2 )+VL2(N6,K-1))
  210 D15(K)=HH1*D2(K)*PT(N6,J,K)+HH4*(D9(K)*PT(NIV,J,K+1)+D9(K-1)*PT
     1 (NIV,J,K-1))+HH2*(D10(K)*PT(NIV,J+1,K)+D11(K)*PT(NIV,J-1,K))+DSU*
     2 (-D6(K)*PT(N6,J+1,K)+D7(K)*PT(N6,J-1,K))+DSV*(-D8(K)*PT(N6,J,K+1)
     3 +D8(K-1)*PT(N6,J,K-1))+DUV*(VL1(N6,K)*PT(NIV,J+1,K+1)-VL1(N6,K-1)
     4 *PT(NIV,J+1,K-1)-VL2(N6,K)*PT(NIV,J-1,K+1)+VL2(N6,K-1)*PT(NIV,J-1
     5 ,K-1))
        DO 220 K=2,N4
        PAS2=PM2*PT(NIV,J,K)+(OM2*D15(K))/D1(K)
        BB=ABS(PAS2-PT(NIV,J,K))/DT
        AA1(5)=AA1(5)+BB*BB
        ERVA=AMAX1(ERVA,BB)
  220 PT(NIV,J,K)=PAS2
        PT(NIV,J,1)=PT(NIV,J,N4)-C2
        PT(NIV,J,N5)=PT(NIV,J,2)+C2
  230 CONTINUE
        IF (NCO.LT.NV) GO TO 310
        DO 250 I=1,N6
        DO 240 K=2,N4
        D2(K)=HH1*(VA1(I,K)+VA2(I,K)+VA1(I,K-1)+VA2(I,K-1))*(PT(I+1,J,K)
     1 -PT(I,J,K))*(PT(I+1,J,K)-PT(I,J,K))+DSU*(VD2(I,K)+VD2(I,K-1))*(
     2 (PT(I+1,J,K)-PT(I,J-1,K))*(PT(I+1,J,K)-PT(I,J-1,K))-(PT(I+1,J-1,K
     3 )-PT(I,J,K))*(PT(I+1,J-1,K)-PT(I,J,K)))+DSV*(VE1(I,K)+VE2(I,K))*(
     4 (PT(I+1,J,K+1)-PT(I,J,K))*(PT(I+1,J,K+1)-PT(I,J,K))-(PT(I+1,J,K)
     5 -PT(I,J,K+1))*(PT(I+1,J,K)-PT(I,J,K+1)))
  240 CONTINUE
        DO 250 K=2,N4
        A11=A11+D2(K)
  250 CONTINUE
        DO 270 I=2,N6
        DO 260 K=2,N4
        D2(K)=HH4*(VB1(I,K)+VB2(I,K)+VB1(I-1,K)+VB2(I-1,K))*(PT(I,J,K+1)
     1 -PT(I,J,K))*(PT(I,J,K+1)-PT(I,J,K))+HH2*(VC2(I,K)+VC2(I,K-1)+VC2
     2 (I-1,K)+VC2(I-1,K-1))*(PT(I,J,K)-PT(I,J-1,K))*(PT(I,J,K)-PT(I,J-1
     3 ,K))+DUV*(VL2(I,K)+VL2(I-1,K))*((PT(I,J,K+1)-PT(I,J-1,K))*(PT(I,J
     4 ,K+1)-PT(I,J-1,K))-(PT(I,J,K)-PT(I,J-1,K+1))*(PT(I,J,K)-PT(I,J-1
     5 ,K+1)))
```

```
  260 CONTINUE
      DO 270 K=2,N4
      A11=A11+D2(K)
  270 CONTINUE
      DO 280 K=2,N4
      D1(K)=HH4*(VB1(1,K)+VB2(1,K))*(PT(1,J,K+1)-PT(1,J,K))*(PT(1,J,K+1)
     1 -PT(1,J,K))+HH2*(VC2(1,K)+VC2(1,K-1))*(PT(1,J,K)-PT(1,J-1,K))*(PT
     2 (1,J,K)-PT(1,J-1,K))+DUV*VL2(1,K)*((PT(1,J,K+1)-PT(1,J-1,K))*(PT(
     3 1,J,K+1)-PT(1,J-1,K))-(PT(1,J,K)-PT(1,J-1,K+1))*(PT(1,J,K)-PT(1,J
     4 -1,K+1)))
  280 CONTINUE
      DO 290 K=2,N4
      D2(K)=HH4*(VB1(N6,K)+VB2(N6,K))*(PT(NIV,J,K+1)-PT(NIV,J,K))*(PT
     1 (NIV,J,K+1)-PT(NIV,J,K))+HH2*(VC2(N6,K)+VC2(N6,K-1))*(PT(NIV,J,K)
     2 -PT(NIV,J-1,K))*(PT(NIV,J,K)-PT(NIV,J-1,K))+DUV*VL2(N6,K)*((PT
     3 (NIV,J,K+1)-PT(NIV,J-1,K))*(PT(NIV,J,K+1)-PT(NIV,J-1,K))-(PT(NIV
     4 ,J,K)-PT(NIV,J-1,K+1))*(PT(NIV,J,K)-PT(NIV,J-1,K+1)))
  290 CONTINUE
      DO 300 K=2,N4
  300 A11=A11+D1(K)+D2(K)
  310 CONTINUE
      IF (NPLOT.LT.0) GO TO 330
C     AVERAGE POLOIDAL AND TOROIDAL VACUUM FIELDS
      DO 320 I=1,N6
      BPV(I)=BPV(I)*HU*HV
  320 BTV(I)=BTV(I)*HU*HV
      GO TO 360
  330 CONTINUE
C     PERIODICITY CONDITIONS
      DO 340 I=1,NIV
      DO 340 K=1,N5
      PT(I,1,K)=PT(I,N1,K)-C1
  340 PT(I,N2,K)=PT(I,2,K)+C1
      IF (NCO.GT.300) GO TO 350
      IF (ERVA.GT.AT) GO TO 30
      IF (NCO.LT.NV) GO TO 30
      HURV=HU*HR*HV
      A11=0.5*A11*HURV
      EVNE=-0.5*A11
      GO TO 360
  350 PRINT 370
      NPLOT=1
      CALL TGRAD (X1,X2,X3,X4,X5,X6,X7)
      CALL FPRINT (ITER-1,-1)
      CALL TPLOT (ITER-1,0)
      STOP
  360 CONTINUE
      RETURN
C
  370 FORMAT (///,6X,34HLAPLACE EQUATION DOES NOT CONVERGE)
      END
```

```
      SUBROUTINE CV (J)
C     EVALUATES COEFFICIENTS OF VACUUM EQUATIONS
      USE NAME1
      USE NAME4
      USE NAME12
      USE NAME13
      DO 10 K=2,N4
      D1(K)=R(J,K)-RB1(K)
   10 D2(K)=Z(J,K)-ZB1(K)
      DO 50 I=1,N6
      S=(I-0.5)*HR
      S1=1.0-S
      DO 20 K=2,N4
      D8(K)=ZLE*(1.0+EP*(RB1(K)+S*D1(K)))
      D3(K)=S1*RBU1(K)+S*RU(J,K)
      D4(K)=S1*ZBU1(K)+S*ZU(J,K)
      D5(K)=S1*RBV1(K)+S*RV(J,K)
      D6(K)=S1*ZBV1(K)+S*ZV(J,K)
   20 CONTINUE
      DO 30 K=2,N4
      VE1(I,K)=(D3(K)*D6(K)-D5(K)*D4(K))/D8(K)
      VL1(I,K)=(D5(K)*D2(K)-D1(K)*D6(K))/D8(K)
   30 VB1(I,K)=(D1(K)*D4(K)-D2(K)*D3(K))/D8(K)
      DO 40 K=2,N4
      VA1(I,K)=(D3(K)*D3(K)+D4(K)*D4(K)+VE1(I,K)*VE1(I,K))/VB1(I,K)
      VC1(I,K)=(D1(K)*D1(K)+D2(K)*D2(K)+VL1(I,K)*VL1(I,K))/VB1(I,K)
   40 VD1(I,K)=(VE1(I,K)*VL1(I,K)-(D3(K)*D1(K)+D4(K)*D2(K)))/VB1(I,K)
   50 CONTINUE
      DO 60 I=1,N6
      VA1(I,1)=VA1(I,N4)
      VB1(I,1)=VB1(I,N4)
      VC1(I,1)=VC1(I,N4)
      VD1(I,1)=VD1(I,N4)
      VL1(I,1)=VL1(I,N4)
   60 VE1(I,1)=VE1(I,N4)
      RETURN
      END




      SUBROUTINE TBO (ERBO,ERBO1,IRQ)
C     COMPUTES ITERATION OF 3D FREE BOUNDARY EQUATION
      USE NAME1
      USE NAME5
      USE NAME7
      USE NAME8
      USE NAME9
      USE NAME10
```

```
       USE NAME12
       USE NAME13
       ERB01=0.0000001
C      COMPUTE PRESSURE AND S DERIVATIVE OF R**2 AT THE BOUNDARY
       PB=PRM(NI)
       ZZ=ZLE*ZLE
       C11=0.0
       C22=1.0
       C12=0.0
       C22=0.0
       B1=1.0/(2.0*HU*HU)
       B2=1.0/(2.0*HV*HV)
       B3=1.0/(4.0*HU*HV)
       B4=1.0/(2.0*HU)
       B5=1.0/(2.0*HV)
       EPL=EP*ZLE
       DD1=0.5*(1.0+QS(N3))
       DD2=0.125*QS(N3)
       DO 160 J=2,N1
       CALL CBO (J)
       DO 10 K=2,N4
       D1(K)=(PT(1,J+1,K)-PT(1,J,K))*(PT(1,J+1,K)-PT(1,J,K))
 10    CONTINUE
       D1(N5)=D1(2)
       DO 20 K=2,N4
       D6(K)=PT(1,J,K+1)-PT(1,J,K)
 20    D7(K)=PT(1,J+1,K+1)-PT(1,J+1,K)
       DO 30 K=2,N4
       D9(K)=PT(1,J+1,K)+PT(1,J+1,K+1)-PT(1,J,K)-PT(1,J,K+1)
 30    D11(K)=PT(1,J,K+1)+PT(1,J+1,K+1)-PT(1,J,K)-PT(1,J+1,K)
       DO 40 K=2,N4
       D1(K)=B1*(D1(K)+D1(K+1))
       D2(K)=B2*(D6(K)*D6(K)+D7(K)*D7(K))
       D3(K)=B3*D9(K)*D11(K)
 40    CONTINUE
       DO 50 K=2,N4
       D4(K)=B1*(AL(NI,J+1,K)-AL(NI,J,K))*(AL(NI,J+1,K)-AL(NI,J,K))
       D5(K)=B2*((AL(NI,J,K+1)-AL(NI,J,K))*(AL(NI,J,K+1)-AL(NI,J,K))+(AL
      1 (NI,J+1,K+1)-AL(NI,J+1,K))*(AL(NI,J+1,K+1)-AL(NI,J+1,K)))
 50    CONTINUE
       D4(N5)=D4(2)
       DO 60 K=2,N4
       D4(K)=D4(K)+D4(K+1)
       D6(K)=B3*(AL(NI,J+1,K)+AL(NI,J+1,K+1)-AL(NI,J,K)-AL(NI,J,K+1))*(AL
      1 (NI,J,K+1)+AL(NI,J+1,K+1)-AL(NI,J,K)-AL(NI,J+1,K))
 60    CONTINUE
       DO 70 K=2,N4
       HB1(K)=(RB1(K)-0.5*(RA(K)+RA(K+1)))*ZBU1(K)-(ZB1(K)-0.5*(ZA(K)+ZA
      1 (K+1)))*RBU1(K)
       D7(K)=RO(N3,J+1,K)*RO(N3,J+1,K)+RO(N3,J,K)*RO(N3,J,K)
       D8(K)=ZLE*(1.0+EP*RB1(K))
 70    D9(K)=(RBU1(K)*ZBV1(K)-RBV1(K)*ZBU1(K))/D8(K)
       D7(N5)=D7(2)
```

```
      DO 80 K=2,N4
      D7(K)=HB1(K)*(DD1-DD2*(D7(K)+D7(K+1)))
      D13(K)=D8(K)*D8(K)
      D10(K)=RBU1(K)*RBU1(K)+ZBU1(K)*ZBU1(K)
   80 CONTINUE
      DO 90 K=2,N4
      D11(K)=(D10(K)*D2(K)+(D13(K)+RBV1(K)*RBV1(K)+ZBV1(K)*ZBV1(K))*D1(K
     1 )-2.0*(RBU1(K)*RBV1(K)+ZBU1(K)*ZBV1(K))*D3(K))/(D13(K)*(D10(K)+D9
     2 (K)*D9(K)))
      D12(K)=(D10(K)*D5(K)+(D13(K)+RBV1(K)*RBV1(K)+ZBV1(K)*ZBV1(K))*D4(K
     1 )-2.0*(RBU1(K)*RBV1(K)+ZBU1(K)*ZBV1(K))*D6(K))/(D13(K)*D7(K)*D7(K
     2 ))
      AK(J,K)=(1.0+EP*RB1(K))*(0.5*(D12(K)-D11(K))+PB)
      D10(K)=D10(K)+D9(K)*D9(K)
   90 CONTINUE
      DO 100 K=2,N4
      D2(K)=(D5(K)/(D7(K)*D7(K))-D2(K)/D10(K))/D13(K)
      D1(K)=(D4(K)/(D7(K)*D7(K))-D1(K)/D10(K))/D13(K)
      D3(K)=(D6(K)/(D7(K)*D7(K))-D3(K)/D10(K))/D13(K)
  100 CONTINUE
      DO 110 K=2,N4
      D4(K)=R(J,K)-0.5*(RVA(K)+RVA(K+1))
      D5(K)=Z(J,K)-0.5*(ZVA(K)+ZVA(K+1))
      D6(K)=D4(K)*RBU1(K)+D5(K)*ZBU1(K)
      D7(K)=D4(K)*RBV1(K)+D5(K)*ZBV1(K)
  110 CONTINUE
      DO 120 K=2,N4
      AKFV(J,K)=(1.0+EP*RB1(K))*(D1(K)*D7(K)-D3(K)*D6(K)-D9(K)*D11(K)*
     1 (D4(K)*ZBU1(K)-D5(K)*RBU1(K))/(D10(K)*D8(K)))
  120 AKFU(J,K)=D2(K)*D6(K)-D3(K)*D7(K)+D11(K)*(D6(K)-D9(K)*(D5(K)*RBV1
     1 (K)-D4(K)*ZBV1(K))/D8(K))/D10(K)
      DO 130 K=2,N4
      D6(K)=0.5*(RVA(K)+RVA(K+1)-RA(K)-RA(K+1))
  130 D7(K)=0.5*(ZVA(K)+ZVA(K+1)-ZA(K)-ZA(K+1))
      DO 140 K=2,N4
      AKFU(J,K)=(1.0+EP*RB1(K))*(AKFU(J,K)-D12(K)*(D6(K)*D5(K)-D7(K)*D4
     1 (K))/HB1(K))
      D6(K)=(D6(K)*ZU(J,K)-D7(K)*RU(J,K)+0.5*(X(J,K)+X(J+1,K)+X(J,K+1)+X
     1 (J+1,K+1))*(D4(K)*ZU(J,K)-D5(K)*RU(J,K)))/HB1(K)
      D7(K)=RV(J,K)-(RVA(K+1)-RVA(K))/HV
  140 D13(K)=ZV(J,K)-(ZVA(K+1)-ZVA(K))/HV
      DO 150 K=2,N4
      PAS7=RBU1(K)*RU(J,K)+ZBU1(K)*ZU(J,K)
      PAS8=RBV1(K)*D7(K)+ZBV1(K)*D13(K)+EPL*D8(K)*D4(K)
      PAS9=RBU1(K)*D7(K)+ZBU1(K)*D13(K)+RBV1(K)*RU(J,K)+ZBV1(K)*ZU(J,K)
      EF=(-EPL*D4(K)*D9(K)-(D7(K)*ZBU1(K)-D13(K)*RBU1(K)+RBV1(K)*ZU(J,K)
     1 -ZBV1(K)*RU(J,K)))/D8(K)
      AKF(J,K)=EP*D4(K)*(PB-0.5*(D12(K)-D11(K)))+(1.0+EP*RB1(K))*(-D12(K
     1 )*D6(K)+D11(K)*(PAS7+D9(K)*EF)/D10(K)+D2(K)*PAS7+D1(K)*PAS8-D3(K)
     2 *PAS9)
  150 CONTINUE
      AK(J,1)=AK(J,N4)
      AKF(J,1)=AKF(J,N4)
```

```
      AKFU(J,1)=AKFU(J,N4)
  160 AKFV(J,1)=AKFV(J,N4)
      DO 170 K=1,N4
      D10(K)=0.0
      D11(K)=0.0
      AK(1,K)=AK(N1,K)
      AKF(1,K)=AKF(N1,K)
      AKFU(1,K)=AKFU(N1,K)
  170 AKFV(1,K)=AKFV(N1,K)
C     COMPUTE LAX-WENDROFF ITERATION
      CALL CBO (1)
      DO 210 J=2,N1
      DO 180 K=1,N4
      RBU2(K)=RBU1(K)
  180 ZBU2(K)=ZBU1(K)
      CALL CBO (J)
      DO 190 K=2,N4
      D1(K)=0.25*(AK(J,K)+AK(J-1,K)+AK(J,K-1)+AK(J-1,K-1))
      D2(K)=0.25*(AKFU(J,K)+AKFU(J-1,K)+AKFU(J,K-1)+AKFU(J-1,K-1))
      D3(K)=0.25*(AKFV(J,K)+AKFV(J-1,K)+AKFV(J,K-1)+AKFV(J-1,K-1))
      D4(K)=B4*(AK(J,K)+AK(J,K-1)-AK(J-1,K)-AK(J-1,K-1))
      D5(K)=B5*(AK(J,K)+AK(J-1,K)-AK(J,K-1)-AK(J-1,K-1))
      D7(K)=0.25*(RBU1(K)+RBU1(K-1)+RBU2(K)+RBU2(K-1))
      D8(K)=0.25*(ZBU1(K)+ZBU1(K-1)+ZBU2(K)+ZBU2(K-1))
  190 D6(K)=0.25*(AKF(J,K)+AKF(J-1,K)+AKF(J,K-1)+AKF(J-1,K-1))
      DO 200 K=2,N4
      X7=D1(K)+0.5*DTB*(D1(K)*D6(K)+D2(K)*D4(K)+D3(K)*D5(K))
      D10(K)=D10(K)+X7*D8(K)*(1.0-X(J,K))*HU
      D11(K)=D11(K)-X7*D7(K)*(1.0-X(J,K))*HU
  200 AKB(J,K)=X7
  210 CONTINUE
      IF (IRQ.GT.0) GO TO 310
      IF (PO1.LT.0.0) GO TO 260
      PO2=0.0
      DO 230 J=2,N1
      X1=0.0
      DO 220 K=2,N4
  220 X1=X1+AKB(J,K)
      X1=X1*HV
  230 PO2=PO2+X1*SU(3,J)
      PO2=PO2*2.0*HU
      DO 250 J=2,N1
      DO 240 K=2,N4
  240 AKB(J,K)=AKB(J,K)-PO2*SU(3,J)
  250 CONTINUE
  260 CONTINUE
      ERBO=0.0
      DO 280 J=2,N1
      DO 270 K=2,N4
      AA=ABS(AKB(J,K))
      ERBO=AMAX1(ERBO,AA)
  270 X(J,K)=X(J,K)+DTB*AKB(J,K)
      X(J,1)=X(J,N4)
```

```
  280 X(J,N5)=X(J,2)
      DO 290 K=1,N5
      X(1,K)=X(N1,K)
  290 X(N2,K)=X(2,K)
      IF (IVAX.LT.0) GO TO 310
      DO 300 K=2,N4
      ERBO1=AMAX1(ERBO1,ABS(D10(K)),ABS(D11(K)))
      RVA(K)=RVA(K)+DTB*D10(K)
  300 ZVA(K)=ZVA(K)+DTB*D11(K)
      RVA(1)=RVA(N4)
      RVA(N5)=RVA(2)
      ZVA(1)=ZVA(N4)
      ZVA(N5)=ZVA(2)
  310 CONTINUE
      RETURN
      END




      SUBROUTINE ELAM
C     COMPUTES DESCENT COEFFICIENT FOR ITERATIVE SCHEME
      USE NAME11
      USE NAME12
      USE NAME13
      NE1=NE-1
      NE2=NE-2
      DT2=2.0*DT
      EN(NE)=0.0
      X1=(OP(2)-OP(1))/(DT*(OP(2)+OP(1)))
      X2=OE(2)+OE(1)
      DO 10 I=2,NE1
      X3=(OP(I+1)-OP(I))/(DT*(OP(I+1)+OP(I)))
      X4=OE(I)+OE(I+1)
      X5=(OE(I+1)-OE(I-1))/DT2
      X6=X1+X3
      EN(NE)=EN(NE)+(X3-X1)/DT+0.25*X6*X6+X6*(X2+X4)/8.0+X5
      X1=X3
   10 X2=X4
      EN(NE)=ABS(EN(NE)/NE2)
      RETURN
      END
```

```
      FUNCTION BNORM (I1,I2,X,F,XX1,XX2)
C     COMPUTES L2 NORM
      DIMENSION X(101), F(101)
      IN=I1
      IEND=I2
      XIN=XX1*XX1
      XEND=XX2*XX2
      BNORM=0.0
      IEND1=IEND-1
      DO 10 I=IN,IEND1
      X1=(X(I+1)-X(I))*0.5*(F(I)*F(I)+F(I+1)*F(I+1))
  10  BNORM=BNORM+X1
      BNORM=BNORM/(XEND-XIN)
      BNORM=SQRT(BNORM)
      RETURN
      END




      SUBROUTINE SPLIF (M,N,S,F,FP,FPP,FPPP,KM,VM,KN,VN,MODE,FQM,IND)
C     GENERAL PURPOSE SPLINE FIT
      DIMENSION S(1), F(1), FP(1), FPP(1), FPPP(1)
      IND=0
      K=IABS(N-M)
      IF (K-1) 180,180,10
  10  K=(N-M)/K
      I=M
      J=M+K
      DS=S(J)-S(I)
      D=DS
      IF (DS) 20,180,20
  20  DF=(F(J)-F(I))/DS
      IF (KM-2) 30,40,50
  30  U=.5
      V=3.*(DF-VM)/DS
      GO TO 80
  40  U=0.
      V=VM
      GO TO 80
  50  U=-1.
      V=-DS*VM
      GO TO 80
  60  I=J
      J=J+K
      DS=S(J)-S(I)
      IF (D*DS) 180,180,70
  70  DF=(F(J)-F(I))/DS
      B=1./(DS+DS+U)
```

```
          U=B*DS
          V=B*(6.*DF-V)
   80 FP(I)=U
          FPP(I)=V
          U=(2.-U)*DS
          V=6.*DF+DS*V
          IF (J-N) 60,90,60
   90 IF (KN-2) 100,110,120
  100 V=(6.*VN-V)/U
          GO TO 130
  110 V=VN
          GO TO 130
  120 V=(DS*VN+FPP(I))/(1.+FP(I))
  130 B=V
          D=DS
  140 DS=S(J)-S(I)
          U=FPP(I)-FP(I)*V
          FPPP(I)=(V-U)/DS
          FPP(I)=U
          FP(I)=(F(J)-F(I))/DS-DS*(V+U+U)/6.
          V=U
          J=I
          I=I-K
          IF (J-M) 140,150,140
  150 I=N-K
          FPPP(N)=FPPP(I)
          FPP(N)=B
          FP(N)=DF+D*(FPP(I)+B+B)/6.
          IND=1
          IF (MODE) 180,180,160
  160 FPPP(J)=FQM
          V=FPP(J)
  170 I=J
          J=J+K
          DS=S(J)-S(I)
          U=FPP(J)
          FPPP(J)=FPPP(I)+.5*DS*(F(I)+F(J)-DS*DS*(U+V)/12.)
          V=U
          IF (J-N) 170,180,170
  180 RETURN
          END




          SUBROUTINE INTPL (MI,NI,SI,FI,M,N,S,F,FP,FPP,FPPP,MODE)
C         GENERAL PURPOSE INTERPOLATION USING TAYLOR SERIES
          DIMENSION SI(1), FI(1), S(1), F(1), FP(1), FPP(1), FPPP(1)
          K=IABS(N-M)
```

```
          K=(N-M)/K
          I=M
          MIN=MI
          NIN=NI
          D=S(N)-S(M)
          IF (D*(SI(NI)-SI(MI))) 10,20,20
   10 MIN=NI
          NIN=MI
   20 KI=IABS(NIN-MIN)
          IF (KI) 40,40,30
   30 KI=(NIN-MIN)/KI
   40 II=MIN-KI
          C=0.
          IF (MODE) 60,60,50
   50 C=1.
   60 II=II+KI
          SS=SI(II)
   70 I=I+K
          IF (I-N) 80,90,80
   80 IF (D*(S(I)-SS)) 70,70,90
   90 J=I
          I=I-K
          SS=SS-S(I)
          FPPPP=C*(FPPP(J)-FPPP(I))/(S(J)-S(I))
          FF=FPPP(I)+.25*SS*FPPPP
          FF=FPP(I)+SS*FF/3.
          FF=FP(I)+.5*SS*FF
          FI(II)=F(I)+SS*FF
          IF (II-NIN) 60,100,60
  100 RETURN
          END




          SUBROUTINE PINT (N,I1,I2,W,X,F,FX,C,SIG)
   C      GENERAL PURPOSE POLYNOMIAL INTERPOLATION
          DIMENSION F(50), X(50), XW(50), A(10,10), B(10), C(10), AW(10,10),
        1 WKS1(10), WKS2(10), XK(50), XIK(50,6), FX(50)
          SIG=0.0
          NP=N+1
          DO 20 I=I1,I2
          XIK(I,1)=1.0
          DO 10 J=2,NP
   10 XIK(I,J)=XIK(I,J-1)*X(I)
   20 XW(I)=ABS(X(I))**W
          DO 50 K=1,NP
          B(K)=0.0
          DO 30 I=I1,I2
```

```
   30 B(K)=B(K)+XW(I)*F(I)*XIK(I,K)
      DO 40 J=1,NP
      A(K,J)=0.0
      DO 40 I=I1,I2
   40 A(K,J)=A(K,J)+XW(I)*XIK(I,K)*XIK(I,J)
   50 CONTINUE
      CALL F04ATF (A,10,B,NP,C,AW,10,WKS1,WKS2,IFAIL)
      IF (IFAIL.EQ.0) GO TO 60
      PRINT 90, IFAIL
   60 SUM=0.0
      DO 80 I=I1,I2
      SUM=SUM+XW(I)
      FX(I)=0.0
      DO 70 K=1,NP
   70 FX(I)=FX(I)+C(K)*XIK(I,K)
      FB=FX(I)-F(I)
   80 SIG=SIG+XW(I)*FB*FB
      SIG=SQRT(SIG/SUM)
      RETURN
C
   90 FORMAT (//,3X,"IFAIL=",I5,//)
      END




      SUBROUTINE ASOUT
C     PRINTS FINAL RESULTS FOR AXIALLY SYMMETRIC SOLUTION
      USE NAME3
      USE NAME7
      USE NAME8
      USE NAME10
      USE NAME12
      USE NAME13
      PRINT 50, RA,ZA
      IF(NVAC.LT.0) GO TO 1
      PRINT 80, R2,Z2
    1 PRINT 60
      PR(1)=PRP(1)
      PR(NI)=PRM(NI)
      DO  2 I=2,N3
    2 PR(I)=0.5*(PRP(I)+PRM(I))
      PRINT 70, (PR(I),I=1,NI)
      PRINT 120, ENER
      IF (NVAC.LT.0) GO TO 30
      PRINT 130, EVAC,ETOT
      DO 20 I=1,7
      XR(I)=0.0
      DO 10 J=2,N1
```

```
   10 XR(I)=XR(I)+X(J)*SU(I,J)
   20 XR(I)=2.0*HU*XR(I)
      XR(1)=0.5*XR(1)
      PRINT 100
      PRINT 110, (XR(I),I=1,7)
   30 CONTINUE
      RETURN
C
   50 FORMAT (//6X,"MAGNETIC AXIS",3X,"R=",F9.5,5X,"Z=",F9.5,/)
   60 FORMAT (6X,8HPRESSURE/)
   70 FORMAT (6X,13F7.3/6X,13F7.3)
   80 FORMAT (6X,"VACUUM AXIS",5X,"R=",F9.5,5X,"Z=",F9.5//)
  100 FORMAT (//6X,38HFOURIER COEFFICIENTS FOR FREE BOUNDARY//)
  110 FORMAT (14X,5HCONST,3X,6HSIN(U),3X,6HCOS(U),2X,7HSIN(2U),2X,7HCOS(
     12U),2X,7HSIN(3U),2X,7HCOS(3U)/10X,7(2X,F7.4))
  120 FORMAT (//6X,"PLASMA ENERGY =",E17.10,/)
  130 FORMAT(6X,"VACUUM ENERGY =",E17.10,//6X,"HAMILTONIAN   =",E17.10,
     1//)
      END




      SUBROUTINE FPRINT (ITER,NEQ)
C     PRINTS FINAL RESULTS FOR 3D SOLUTION
      USE NAME1
      COMMON /AUX/ SP(100), SD(200), SI(200), SJ(200), SIV(50), SJV(50),
     1 SPP(50), SVV(50), F(100), FP(200), FPP(200), FPPP(200), XOR(4),
     2 YOR(4), YB(4,4), SLRV1(50), SIRV1(50), SARV1(50), SLRV2(50),
     3 SIRV2(50), SARV2(50), SLZV1(50), SIZV1(50), SAZV1(50), SLZV2(50),
     4 SIZV2(50), SAZV2(50), SLRA1(50), SIRA1(50), SARA1(50), SLRA2(50),
     5 SIRA2(50), SARA2(50), SLZA1(50), SIZA1(50), SAZA1(50), SLZA2(50),
     6 SIZA2(50), SAZA2(50), SLMK1(50), SIMK1(50), SAMK1(50), SLMK2(50),
     7 SIMK2(50), SAMK2(50), SLMK3(50), SIMK3(50), SAMK3(50), SLMK4(50),
     8 SIMK4(50), SAMK4(50), NITER(50),CONF(11)
      USE NAME7
      USE NAME8
      USE NAME9
      USE NAME10
      USE NAME12
      USE NAME13
      USE NAME14
      USE NAME15
      USE NAME21
      COMMON /CDEN/ DMIN(101), DMAX(101), RPMAX(101), RPMIN(101), CNORP(
     1 101), CNORT(101), BMAX(101), BMIN(101), ABB(101), AJJ(101), ABJ(1
     2 01), AZZ(101), ABZ(101), AW(101), AITP(101), BRIP(101), AMER(101)
     3 , AMERS(101), ASH(101), ASC(101), PLAM1(101), PLAM2(101), RTMAX(1
     4 01), RTMIN(101), TLAM(101), TLAM1(101), TLAM2(101), PLAM(101)
```

```
      5, FACP(101),ALP(101)
        DIMENSION CNOR(101), COMU(50), COVP(50), FMU(50), FVP(50)
        ZERVAL(XE,YE,ZE,UE,VE,WE)=(UE*YE*ZE*(ZE-YE)+VE*XE*ZE*(XE-ZE)+WE*XE
      1 *YE*(YE-XE))/(XE*YE*(YE-XE)+YE*ZE*(ZE-YE)+XE*ZE*(XE-ZE))
        VALZD(XE,YE,UE,VE)=(UE*YE*YE-VE*XE*XE)/(YE*YE-XE*XE)
C         CHOICES FOR NRA AND NZA FROM INPUT CARD 34
        RNAME(1)=8HR,CONST
        RNAME(2)=8HR,SINV
        RNAME(3)=8HR,COSV
        RNAME(4)=8HR,SIN2V
        RNAME(5)=8HR,COS2V
        RNAME(6)=8HR,SIN3V
        RNAME(7)=8HR,COS3V
        ZNAME(1)=8HZ,CONST
        ZNAME(2)=8HZ,SINV
        ZNAME(3)=8HZ,COSV
        ZNAME(4)=8HZ,SIN2V
        ZNAME(5)=8HZ,COS2V
        ZNAME(6)=8HZ,SIN3V
        ZNAME(7)=8HZ,COS3V
C         PRINT OUT RESULTS FOR LAST TIME STEP
        DT1=DT*IC
        IF (NT.LT.0.AND.NEQ.GT.0) GO TO 10
        PRINT 750, ENER
        IF (NVAC.LT.0) GO TO 20
        ABEVAC=ABS(EVAC)
        PRINT 760, ABEVAC,ETOT
        GO TO 20
   10 PRINT 450, ENER,DELTE
        IF (NVAC.LT.0) PRINT 470, SNORM,EVAL
        IF (NVAC.GT.0) PRINT 460, EVAC,SNORM,ETOT,EVAL
   20 PRINT 770
        PRINT 780
        PRINT 790, (XR(I),I=1,7)
        PRINT 800, (XZ(I),I=1,7)
        IF (NVAC.LT.0) GO TO 30
        PRINT 810
        PRINT 780
        PRINT 790, (YR(I),I=1,7)
        PRINT 800, (YZ(I),I=1,7)
        PRINT 820
        GO TO 40
   30 PRINT 830, I1,SNL2(I1)
   40 PRINT 780
        PRINT 840, (SRO(1,L),L=1,7)
        PRINT 850, (SRO(2,L),L=1,7)
        PRINT 860, (SRO(3,L),L=1,7)
        PRINT 870, (SRO(4,L),L=1,7)
        PRINT 880, (SRO(5,L),L=1,7)
        PRINT 890, (SRO(6,L),L=1,7)
        PRINT 900, (SRO(7,L),L=1,7)
C         COMPUTE AVERAGE RADIUS FOR EACH FLUX SURFACE
        DO 60 I=1,NI
```

```
       RP(I)=0.0
       DO 50 J=2,N1
       DO 50 K=2,N4
   50  RP(I)=RP(I)+RO(I,J,K)
       RP(I)=RP(I)*HU*HV*SL1(I)
       IF (NVAC.GT.0) RP(I)=RP(I)*SRO(1,1)
   60  CONTINUE
       IF (NVAC.LT.0) GO TO 80
       DO 70 I=1,NIV
       DS=(I-1)*HR
   70  RM(I)=SRO(1,1)+DS*(1.0-SRO(1,1))
   80  CONTINUE
       IF (NT.LT.0 .AND. NEQ.GT. 0) GO TO 171
       PRINT 910
       PRINT 920
       IF (IABS(NVAC).GT.10) Q(1)=(SL2(3)*Q(2)-Q(3)*SL2(2))/(SL2(3)-SL2(2
      1 ))
       DO 90 I=1,NI
       QQ(I)=Q(I)/QT(I)
       QQ(I)=QQ(I)+IROT
   90  RP1(I)=RP(I)
C      PRINT AVERAGE VALUES
       DO 110 I=1,N3
       SJ(I)=-0.5*(SPVOL(I)/QT(I)+SPM(I+1)/QT(I+1))
  110  CONTINUE
       DO 120 I=2,N3
       SI(I)=0.5*(SJ(I)+SJ(I-1))
       X1=((SPVOL(I)+SPM(I+1))-(SPM(I)+SPVOL(I-1)))/(SL2(I+1)-SL2(I-1))
       X2=0.25*(SPVOL(I)+SPM(I+1)+SPM(I)+SPVOL(I-1))
       PPRIM(I)=GAM*PR(I)*(AMS(I)/AM(I)-X1/X2)
  120  SPPVOL(I)=-(SJ(I)-SJ(I-1))/(0.5*QT(I)*(SL2(I+1)-SL2(I-1)))
       SI(1)=-SPVOL(1)/QT(1)
       SI(NI)=-SPM(NI)/QT(NI)
       SPPVOL(1)=0.0
       SPPVOL(NI)=0.0
       BETA=0.0
       BETA1=0.0
       DO 130 I=1,NI
       SPVOL(I)=SPPVOL(I)/SI(I)
       BETA1=BETA1+BET(I)*PQS(I)
       BETA=BETA+PR(I)*PQS(I)
       PRINT 930, RP1(I),BTP(I),BPP(I),TC(I),PC(I),QQ(I),PR(I),BET(I)
      1,PPRIM(I)
  130  CONTINUE
       BETA=BETA*HS
       BETA1=BETA1*HS
       BETA=2.0*BETA/(BTP(NI)*BTP(NI)+BPP(NI)*BPP(NI))
       DO 140 I=1,50
       COMU(I)=0.0
  140  COVP(I)=0.0
       NC=2
       NC1=NC+1
       CALL PINT (NC,2,NI,0.5,SL2,QQ,FMU,COMU,SIGMU)
```

```
      CALL PINT (NC,2,NI,0.5,SL2,SI,FVP,COVP,SIGVP)
      WD=(COVP(1)-FVP(NI))/COVP(1)
      PRINT 480, BETA1,WD,IROT
      IF (NVAC.LT.0) GO TO 170
      PRINT 940
      PRINT 950, C1,C2,VERT
      PRINT 960
      DO 150 I=1,N6
  150 RV1(I)=0.5*(RM(I)+RM(I+1))
      DO 160 I=1,N6
  160 PRINT 970, RV1(I),BTV(I),BPV(I)
  170 CONTINUE
  171 CONTINUE
      IF (NEQ.LT.0) GO TO 220
      IF (NT.GT.0) PRINT 490
      IF (NT.LT.0) PRINT 500
      PRINT 510
      PRINT 520
      DO 180 I=1,NI
  180 PRINT 530, I,(XL(I,L,1),L=1,7),(XL(I,L,2),L=1,7)
      PRINT 540
      PRINT 520
      DO 190 I=1,NI
  190 PRINT 530, I,(XL(I,L,3),L=1,7),(XL(I,L,4),L=1,7)
      PRINT 550
      PRINT 520
      IF (NT.LT. 0) GO TO 211
      DO 200 I=1,NI
  200 PRINT 530, I,(XO(I,L,1),L=1,7),(XO(I,L,2),L=1,7)
      PRINT 560
      PRINT 520
      DO 210 I=1,NI
  210 PRINT 530, I,(XO(I,L,3),L=1,7),(XO(I,L,4),L=1,7)
      GO TO 220
  211 DO 212 I=1,NI
  212 PRINT 530,I,(-XO(I,L,1),L=1,7),(-XO(I,L,2),L=1,7)
      PRINT 560
      PRINT 520
      DO 213 I=1,NI
  213 PRINT 530, I,(-XO(I,L,3),L=1,7),(-XO(I,L,4),L=1,7)
  220 CONTINUE
      ICD1=(NI-1)/2+1
      IF (NT.LT.0.AND.NEQ.GT.0) GO TO 350
      CALL CDEN (CNOR,ICD1)
      ISTEP=NI/4
      IINI=ICD1-ISTEP
      IEND=IINI+ISTEP
      PRINT 570
      PRINT 580
      N33=N3-1
      DO 240 I=1,NI
      X1=SNL2(I)
      IF (I.EQ.1.OR.I.EQ.NI) GO TO 230
```

```
      PRINT 600, X1,CNORT(I),RTMAX(I),RTMIN(I),DMAX(I),DMIN(I)
      GO TO 240
  230 PRINT 590, X1,DMAX(I),DMIN(I)
  240 CONTINUE
      PRINT 610
      BNORT1=BNORM(3,N33,SL2,CNORT,SL1(3),SL1(N33))*RBOU
      BNORP1=BNORM(3,N33,SL2,CNOR,SL1(3),SL1(N33))
      PRINT 630
      PRINT 620
      DO 250 I=2,N3
      X1=SNL2(I)
  250 PRINT 640, X1,TLAM(I),TLAM1(I),TLAM2(I),PLAM(I),PLAM1(I),PLAM2(I)
      X1=SQRT(2.0)*4.0
      ANB=BNORM(3,N33,SL2,TLAM1,SL1(3),SL1(N33))*RBOU/X1
      PRINT 650
      ANLB=BNORM(3,N33,SL2,TLAM2,SL1(3),SL1(N33))*RBOU/X1
      PRINT 660, BNORT1,MIS,NIS,ANB,MIS,NIS,ANLB
      ANB1=BNORM(3,N33,SL2,PLAM1,SL1(3),SL1(N33))/X1
      ANLB1=BNORM(3,N33,SL2,PLAM2,SL1(3),SL1(N33))/X1
      SLAB=SNL2(IPF)
C     PRINT FOURIER COEFFICIENTS FOR THE PARALLEL CURRENT AND 1/(B.B)
      PRINT 672, IPF,SLAB
      DO 253 M=1,4
      DO 253 N=1,4
      SRO(2*M-1,2*N-1)=ABL(M,N,IPF)
      SRO(2*M-1,2*N)=BBL(M,N+1,IPF)
      SRO(2*M,2*N-1)=CBL(M+1,N,IPF)
  253 SRO(2*M,2*N)=DBL(M+1,N+1,IPF)
      PRINT 781
      PRINT 841, (SRO(1,L),L=1,7)
      PRINT 851, (SRO(2,L),L=1,7)
      PRINT 861, (SRO(3,L),L=1,7)
      PRINT 871, (SRO(4,L),L=1,7)
      PRINT 881, (SRO(5,L),L=1,7)
      PRINT 891, (SRO(6,L),L=1,7)
      PRINT 901, (SRO(7,L),L=1,7)
      PRINT 680, IPF,SLAB
      DO 254 M=1,4
      DO 254 N=1,4
      SRO(2*M-1,2*N-1)=ABI(M,N,IPF)
      SRO(2*M-1,2*N)=BBI(M,N+1,IPF)
      SRO(2*M,2*N-1)=CBI(M+1,N,IPF)
  254 SRO(2*M,2*N)=DBI(M+1,N+1,IPF)
      PRINT 781
      PRINT 841, (SRO(1,L),L=1,7)
      PRINT 851, (SRO(2,L),L=1,7)
      PRINT 861, (SRO(3,L),L=1,7)
      PRINT 871, (SRO(4,L),L=1,7)
      PRINT 881, (SRO(5,L),L=1,7)
      PRINT 891, (SRO(6,L),L=1,7)
      PRINT 901, (SRO(7,L),L=1,7)
C     GEOMETRIC CONFINEMENT TIME
      DO 256 J=1,1
```

```
         DO 255 I=2,N3
  255 ALP(I)=1.0/(SQRT(SNLP(I,J))+0.000001)
         CONF(J)=BNORM(2,N33,SL2,ALP,SL1(2),SL1(N33))*RBOU
  256 CONF(J)=CONF(J)*CONF(J)*100.0
C        PRINT MERCIER CRITERION RESULTS
         N33=N3-1
         DO 320 I=3,N3
         X1=SNL2(I)
         SD(I)=X1
  320 CONTINUE
         DO 330 I=1,21
  330 SP(I)=(I-1)/20.0
         CALL SPLIF (3,N3,SD,ABJ,FP,FPP,FPPP,3,0.0,3,0.0,0,0.0,IND)
         CALL INTPL (1,21,SP,SLMK3,3,N3,SD,ABJ,FP,FPP,FPPP,0)
         PRINT 690
         PRINT 701
         SIMK3(2)=PI*SPVOL(2)
         PRINT 710, SNL2(2),AMERN(2),ASH(2),ABZN(2),AWN(2),ABJN(2)
     1,SIMK3(2),BRIP(2),BS2(2)
         DO 331 I=3,N3
         SIMK3(I)=PI*SPVOL(I)
  331 PRINT 711,SNL2(I),AMERN(I),ASH(I),ABZN(I),AWN(I),ABJN(I),ABJ(I),
     1SIMK3(I),BRIP(I),BS2(I)
         CALL SPLIF (3,N3,SD,AMERN,FP,FPP,FPPP,3,0.0,3,0.0,0,0.0,IND)
         CALL INTPL (1,21,SP,SI,3,N3,SD,AMERN,FP,FPP,FPPP,0)
         CALL SPLIF (3,N3,SD,SPVOL,FP,FPP,FPPP,3,0.0,3,0.0,0,0.0,IND)
         CALL INTPL (1,21,SP,SJ,3,N3,SD,SPVOL,FP,FPP,FPPP,0)
         CALL SPLIF (3,N3,SD,ASH,FP,FPP,FPPP,3,0.0,3,0.0,0,0.0,IND)
         CALL INTPL (1,21,SP,SLMK1,3,N3,SD,ASH,FP,FPP,FPPP,0)
         CALL SPLIF (3,N3,SD,AWN,FP,FPP,FPPP,3,0.0,3,0.0,0,0.0,IND)
         CALL INTPL (1,21,SP,SLMK2,3,N3,SD,AWN,FP,FPP,FPPP,0)
         CALL SPLIF (3,N3,SD,ABZN,FP,FPP,FPPP,3,0.0,3,0.0,0,0.0,IND)
         CALL INTPL (1,21,SP,SIMK1,3,N3,SD,ABZN,FP,FPP,FPPP,0)
         CALL SPLIF (3,N3,SD,ABJN,FP,FPP,FPPP,3,0.0,3,0.0,0,0.0,IND)
         CALL INTPL (1,21,SP,SIMK2,3,N3,SD,ABJN,FP,FPP,FPPP,0)
         CALL SPLIF (3,N3,SD,BRIP,FP,FPP,FPPP,3,0.0,3,0.0,0,0.0,IND)
         CALL INTPL (1,21,SP,SIV,3,N3,SD,BRIP,FP,FPP,FPPP,0)
         PRINT 720
         PRINT 730
         DO 340 I=1,21
         SJ(I)=PI*SJ(I)
         IF ((SP(I).LT.SD(3)).OR.(SP(I).GT.SD(N3))) GO TO 340
         PRINT 740, SP(I),SI(I),SLMK1(I),SIMK1(I),SLMK2(I),SIMK2(I),
     1SLMK3(I),SJ(I),SIV(I)
  340 CONTINUE
         IF (NEQ.LT.0) GO TO 440
  350 CONTINUE
         REWIND 3
         REWIND 4
         XY=FLOAT(ITER)/FLOAT(IC)
         N=INT(XY-1.0)
         NEV=50
         NEV1=MINO(NEV,N)
```

```
      M1=MK1/10
      M2=MK2/10
      M3=MK3/10
      M4=MK4/10
      K1=MK1-M1*10
      K2=MK2-M2*10
      K3=MK3-M3*10
      K4=MK4-M4*10
      XY=FLOAT(N)/50.0
      NN=INT(XY)+1
      NTER=0
      PRINT 1010
      PRINT 980
      PRINT 990
      JJ=0
      JN=0
      DO 360 I=1,N
      NTER=NTER+1
      READ (3) ETOT,ERO,EAL,EAX
      READ (3) (XR(M),M=1,7),(XZ(M),M=1,7)
      READ (3) ((YB(J,K),K=1,4),J=1,4)
      READ (3) EA1,EA2,EA3
      IF (NTER.LT.NN) GO TO 360
      NTER=0
      TT=(I-1)*DT*IC
C     PRINT FOURIER COEFFICIENTS FOR MAGNETIC AXIS
      PRINT 1000, TT,(XR(M),M=1,7),(XZ(M),M=1,7)
  360 CONTINUE
      NTER=0
      REWIND 3
      IF (NVAC.LT.0) GO TO 370
      PRINT 1020
      GO TO 380
  370 PRINT 1030, I1,SNL2(I1)
  380 PRINT 1040
      JJ=0
      JN=0
      DO 390 I=1,N
      NTER=NTER+1
      READ (3) ETOT,ERO,EAL,EAX
      READ (3) (XR(M),M=1,7),(XZ(M),M=1,7)
      READ (3) ((YB(J,K),J=1,4),K=1,4)
      READ (3) EA1,EA2,EA3
      IF (NTER.LT.NN) GO TO 390
      NTER=0
      TT=(I-1)*DT*IC
C     PRINT FOURIER COEFFICIENTS FOR FLUX SURFACE
      PRINT 1050, TT,((YB(J,K),J=1,4),K=1,4)
  390 CONTINUE
      REWIND 3
      REWIND 4
      NTER=0
  400 PRINT 1080
```

```
          PRINT 1090,CONF(1)
    410 DO 430 I=1,N
          NTER=NTER+1
          READ (3) ETOT,ERO,EAL,EAX
          READ (3) (XR(M),M=1,7),(XZ(M),M=1,7)
          READ (3) ((YB(J,K),K=1,4),J=1,4)
          READ (3) EA1,EA2,EA3
          IF (NVAC.LT.0) GO TO 420
          READ (4) ERBO,ERBO1,ERVA
          READ (4) (YR(M),M=1,7),(YZ(M),M=1,7)
          READ (4) OM1,OM2
    420 IF (NTER.LT.NN) GO TO 430
C         PRINT DESCENT COEFFICIENTS
          NTER=0
          TT=(I-1)*DT*IC
          PRINT 1100, TT,EA1
    430 CONTINUE
    440 CONTINUE
          RETURN
C
    450 FORMAT (1H1///6X,10H3D RESULTS///6X,14HPLASMA ENERGY=,E16.9,8X,13H
         1DELTA ENERGY=,E12.5)
    460 FORMAT (/,6X,14HVACUUM ENERGY=,E16.9,8X,13HNORM SQUARED=,E12.5//6X
         1 ,14HTOTAL ENERGY=,E16.9,8X,13HEIGENVALUE  =,E12.5)
    470 FORMAT (/,44X,13HNORM SQUARED=,E12.5//44X,13HEIGENVALUE  =,E12.5)
    480 FORMAT (//,10X,"AVERAGE BETA=",1(1X,F6.3),6X,"WELL DEPTH=",1X,F6.3
         1,5X,"IROT=",I3)
    490 FORMAT (///,50X,32HFOURIER COEFFICIENTS EQUILIBRIUM/)
    500 FORMAT (///,50X,34HFOURIER COEFFICIENTS EIGENFUNCTION/)
    510 FORMAT (36X,7HR CONST,52X,8HR SIN(V)/)
    520 FORMAT (8X,1HI,5X,5HCONST,3X,5HSI(U),3X,5HCO(U),2X,6HSI(2U),2X,6HC
         1O(2U),2X,6HSI(3U),2X,6HCO(3U),3X,5HCONST,3X,5HSI(U),3X,5HCO(U),2X,
         2 6HSI(2U),2X,6HCO(2U),2X,6HSI(3U),2X,6HCO(3U)/)
    530 FORMAT (6X,I3,2X,14F8.4)
    540 FORMAT (//,35X,8HR COS(V),52X,9HR SIN(2V)/)
    550 FORMAT (///,34X,9HPSI CONST,50X,10HPSI SIN(V)/)
    560 FORMAT (//,33X,10HPSI COS(V),50X,11HPSI SIN(2V)/)
    570 FORMAT (//,32X,"J.B/B.B ",21X,"JACOBIAN",/)
    580 FORMAT (10X,"FLUX",9X,"NORM",7X,"MAX     ",3X,"MIN     ",3X,"MAX JAC
         1",3X,"MIN JAC",/)
    590 FORMAT (9X,F5.2,37X,2F10.3)
    600 FORMAT (9X,F5.2,3X,3F10.3,4X,2F10.3)
    610 FORMAT (//,40X,"FOURIER COEFFICIENTS FOR J.B/B.B",//)
    611 FORMAT (//,40X,"FOURIER COEFFICIENTS FOR J.B/B.B.P'",//)
    620 FORMAT (10X,"FLUX",10X,"CONST",3X,"COS(MU-NV)",3X,"SIN(MU-NV)",12X
         1,"CONST",4X,"COS(MU-NV)",3X,"SIN(MU-NV)",8X,/)
    630 FORMAT (/,31X,"J.B/B.B FROM CURL",26X,"J.B/B.B FROM SERIES    ",/)
    640 FORMAT (9X,F5.2,5X,3(F10.4,3X),5X,6(F10.4,3X))
    650 FORMAT (//,35X,"ISLAND WIDTH NORMS",/)
    660 FORMAT (10X,"D(J.B/B.B)=",F8.5,6X,"D(COS",2I1,")=",F8.5,6X,"D(SIN"
         1 ,2I1,")=",F8.5,///)
    671 FORMAT(//,10X,"FOURIER COEFFICIENTS J.B/B.B AT I=",I3,", S="
         1,F5.2," VERSUS PSI AND PHI",//)
```

```
672 FORMAT(//,10X,"FOURIER COEFFICIENTS OF J.B/B.B.P' AT I=",I3,", S="
   1,F5.2," VERSUS PSI AND PHI",//)
680 FORMAT (//,10X,"FOURIER COEFFICIENTS OF 1/B.B AT I=",I3,", S=",F5.
   12," VERSUS PSI AND PHI",//)
690 FORMAT (//,33X,"RESULTS FOR THE MERCIER CRITERION",//)
700 FORMAT (10X,"FLUX",5X,"MERCIER",6X,"ISLAND",6X,"SHEAR2",7X,"JDOTB"
   1,8X,"WELL",6X,"VPP/VP",6X,"MIRROR",//)
701 FORMAT (10X,"FLUX",5X,"MERCIER",6X,"SHEAR2",7X,"JDOTB",8X,"WELL",
   15X,"CURRENT",6X,"ISLAND",6X,"VPP/VP",6X,"MIRROR",4X,"PARSEVAL",//)
710 FORMAT (7X,F7.3,F12.5,4F12.5,12X,3F12.5)
711 FORMAT (7X,F7.3,F12.5,4F12.5,F12.5,3F12.5)
720 FORMAT (//,30X,"SPLINE INTERPOLATED VALUES OF MERCIER RESULTS",//)
730 FORMAT(10X,"FLUX",5X,"MERCIER",6X,"SHEAR2",7X,"JDOTB",8X,"WELL",
   15X,"CURRENT",6X,"ISLAND",6X,"VPP/VP",6X,"MIRROR",//)
740 FORMAT (7X,F7.2,8F12.5)
750 FORMAT (1H1///6X,10H3D RESULTS///6X,14HPLASMA ENERGY=E16.9)
760 FORMAT (/6X,14HVACUUM ENERGY=E16.9//6X,14HHAMILTONIAN  =E16.9)
770 FORMAT (///10X,37HFOURIER COEFFICIENTS OF MAGNETIC AXIS//)
780 FORMAT (25X,5HCONST,5X,6HSIN(V),5X,6HCOS(V),4X,7HSIN(2V),4X,7HCOS(
   12V),4X,7HSIN(3V),4X,7HCOS(3V)/)
781 FORMAT(26X,5HCONST,3X,8HSIN(PHI),3X,8HCOS(PHI),2X,9HSIN(2PHI),2X
   1,9HCOS(2PHI),2X,9HSIN(3PHI),2X,9HCOS(3PHI)/)
790 FORMAT (16X,1HR2X,7F11.5,/)
800 FORMAT (16X,1HZ,2X,7F11.5,/)
810 FORMAT (//10X,32HFOURIER COEFFICIENTS VACUUM AXIS//)
820 FORMAT (//,10X,34HFOURIER COEFFICIENTS FREE BOUNDARY//)
830 FORMAT (///10X,42HFOURIER COEFFICIENTS OF FLUX SURFACE AT I=I3,",
   1="F5.2//)
840 FORMAT (13X,5HCONST,1X,7F11.5/)
841 FORMAT(14X,5HCONST,1X,7F11.5/)
850 FORMAT (12X,6HSIN(U),1X,7F11.5/)
851 FORMAT(11X,8HSIN(PSI),1X,7F11.5/)
860 FORMAT (12X,6HCOS(U),1X,7F11.5/)
861 FORMAT(11X,8HCOS(PSI),1X,7F11.5/)
870 FORMAT (11X,7HSIN(2U)1X,7F11.5/)
871 FORMAT(10X,9HSIN(2PSI),1X,7F11.5/)
880 FORMAT (11X,7HCOS(2U)1X,7F11.5/)
881 FORMAT(10X,9HCOS(2PSI),1X,7F11.5/)
890 FORMAT (11X,7HSIN(3U)1X,7F11.5/)
891 FORMAT(10X,9HSIN(3PSI),1X,7F11.5/)
900 FORMAT (11X,7HCOS(3U)1X,7F11.5/)
901 FORMAT(10X,9HCOS(3PSI),1X,7F11.5/)
910 FORMAT (1H1///15X,55HFLUX SURFACE AVERAGE PLASMA PARAMETERS PER FI
   1ELD PERIOD//)
920 FORMAT (8X,6HRADIUS,5X,4HBTOR,5X,4HBPOL,4X,5HTCURR,4X,5HPCURR,6X,2
   1 HMU,8X,1HP,6X,4HBETA,3X,"PPRIME",/)
930 FORMAT (5X,9F9.3)
940 FORMAT (////6X,13HVACUUM REGION//)
950 FORMAT (10X,17HTOROIDAL CURRENT=F8.3,3X,17HPOLOIDAL CURRENT=F8.3,
   13X,15HVERTICAL FIELD=F8.4)
960 FORMAT (//8X,6HRADIUS,5X,4HBTOR,5X,4HBPOL/)
970 FORMAT (5X,3F9.3)
980 FORMAT (39X,1HR,55X,1HZ/)
```

```
  990 FORMAT (7X,4HTIME,3X,5HCONST,3X,5HSI(V),3X,5HCO(V),2X,6HSI(2V),2X,
     1 6HCO(2V),2X,6HSI(3V),2X,6HCO(3V),3X,5HCONST,3X,5HSI(V),3X,5HCO(V)
     2 ,2X,6HSI(2V),2X,6HCO(2V),2X,6HSI(3V),2X,6HCO(3V)/)
 1000 FORMAT (5X,F6.2,14F8.4)
 1010 FORMAT (///,50X,34HFOURIER COEFFICIENTS MAGNETIC AXIS/)
 1020 FORMAT (///,50X,34HFOURIER COEFFICIENTS FREE BOUNDARY/)
 1030 FORMAT (///,50X,33HFOURIER COEFFICIENTS RADIUS AT I=I3,5H , S=F5.2
     1 /)
 1040 FORMAT (7X,4HTIME,2X,5HMK=00,2X,5HMK=10,2X,5HMK=20,2X,5HMK=30,2X,5
     1 HMK=01,2X,5HMK=11,2X,5HMK=21,2X,5HMK=31,2X,5HMK=02,2X,5HMK=12,2X,
     2 5HMK=22,2X,5HMK=32,2X,5HMK=03,2X,5HMK=13,2X,5HMK=23,2X,5HMK=33,/)
 1050 FORMAT (5X,F6.2,16F7.4)
 1080 FORMAT (1H1///17X,19HDESCENT COEFFICIENT,10X,"GEOMETRIC CONFINEMEN
     1T TIME",/)
 1090 FORMAT (18X,4HTIME,4X,5HE1/A1,24X,"TAU=",F6.2,/)
 1100 FORMAT (16X,F6.2,2X,F8.3)
      END



      SUBROUTINE PRNT
C     CHOOSES DATA TO BE PRINTED ACCORDING TO INPUT ON CARD 26
      COMMON /PRINT/ IX(8), IJ(8), JX(50), PJX(50), NNJ, NJX
      JX(1)=8HAXIS ERR
      JX(2)=8H  RO ERR
      JX(3)=8H PSI ERR
      JX(4)=8H BOU ERR
      JX(5)=8H VAC ERR
      JX(6)=8H DELENER
      JX(7)=8H JAC RAT
      JX(8)=8HROAX ERR
      JX(9)=8HRMA CONS
      JX(10)=8H RMA SIV
      JX(11)=8H RMA COV
      JX(12)=8HRMA SI2V
      JX(13)=8HRMA CO2V
      JX(14)=8HRMA SI3V
      JX(15)=8HRMA CO3V
      JX(16)=8HZMA CONS
      JX(17)=8H ZMA SIV
      JX(18)=8H ZMA COV
      JX(19)=8HZMA SI2V
      JX(20)=8HZMA CO2V
      JX(21)=8HZMA SI3V
      JX(22)=8HZMA CO3V
      JX(23)=8H    MK=00
      JX(24)=8H    MK=10
      JX(25)=8H    MK=20
```

```
      JX(26)=8H    MK=30
      JX(27)=8H    MK=01
      JX(28)=8H    MK=11
      JX(29)=8H    MK=21
      JX(30)=8H    MK=31
      JX(31)=8H    MK=02
      JX(32)=8H    MK=12
      JX(33)=8H    MK=22
      JX(34)=8H    MK=32
      JX(35)=8H    MK=03
      JX(36)=8H    MK=13
      JX(37)=8H    MK=23
      JX(38)=8H    MK=33
      JX(39)=8H       E1
      JX(40)=8H       E2
      JX(41)=8H       E3
      JX(42)=8H      OM1
      JX(43)=8H      OM2
      JX(44)=8H   MU ERR
      JX(45)=8H      A11
      JX(46)=8H      A22
      JX(47)=8H      A12
      JX(48)=8H     VERT
      JX(49)=8H   ENERGY
      JX(50)=8H     FPSI
      NNJ=0
      NJX=50
      DO 20 J=1,8
      DO 10 K=1,NJX
      IF (IX(J).NE.JX(K)) GO TO 10
      NNJ=NNJ+1
      IJ(NNJ)=K
   10 CONTINUE
   20 CONTINUE
      RETURN
      END




      SUBROUTINE TPLOT (ITER,NISL)
C     PLOTS FINAL RESULTS
      USE NAME1
      USE NAME14
      COMMON /AUX/ SP(100), SD(200), SI(200), SJ(200), SIV(50), SJV(50),
     1 SPP(50), SVV(50), F(100), FP(200), FPP(200), FPPP(200), XOR(4),
     2 YOR(4), YB(4,4), SLRV1(50), SIRV1(50), SARV1(50), SLRV2(50),
     3 SIRV2(50), SARV2(50), SLZV1(50), SIZV1(50), SAZV1(50), SLZV2(50),
     4 SIZV2(50), SAZV2(50), SLRA1(50), SIRA1(50), SARA1(50), SLRA2(50),
```

```
     5 SIRA2(50), SARA2(50), SLZA1(50), SIZA1(50), SAZA1(50), SLZA2(50),
     6 SIZA2(50), SAZA2(50), SLMK1(50), SIMK1(50), SAMK1(50), SLMK2(50),
     7 SIMK2(50), SAMK2(50), SLMK3(50), SIMK3(50), SAMK3(50), SLMK4(50),
     8 SIMK4(50), SAMK4(50), NITER(50)
       USE NAME7
       USE NAME8
       USE NAME9
       USE NAME10
       USE NAME11
       USE NAME12
       USE NAME13
       USE NAME15
       USE NAME21
       COMMON /CDEN/ DMIN(101), DMAX(101), RPMAX(101), RPMIN(101), CNORP(
     1 101), CNORT(101), BMAX(101), BMIN(101), ABB(101), AJJ(101), ABJ(1
     2 01), AZZ(101), ABZ(101), AW(101), AITP(101), BRIP(101), AMER(101)
     3 , AMERS(101), ASH(101), ASC(101), PLAM1(101), PLAM2(101), RTMAX(1
     4 01), RTMIN(101), TLAM(101), TLAM1(101), TLAM2(101), PLAM(101)
     5, FACP(101),ALP(101)
       DIMENSION PL1(1500), PL2(1500), PL3(1500), PL4(1500), PL5(1500)
       DIMENSION XPL(1500,5)
       EQUIVALENCE (XO(1,1,1),XPL(1,1))
       EQUIVALENCE (XPL(1,1),PL1(1)), (XPL(1,2),PL2(1)), (XPL(1,3),PL3(1)
     1 ), (XPL(1,4),PL4(1)), (XPL(1,5),PL5(1))
       DIMENSION TEXTX(3), TEXTY(3), SYM(4), T(7), THE(4)
C      INITIALIZE PLOT PACKAGE
       DO 10 I=1,7
    10 T(I)=0.
       ENCODE (27,570,T) NRUN
       CALL KEEP80 (1,3)
       CALL FR80ID
       CALL MAP (0.0,12.0,0.0,12.0)
       GO TO 190
    20 CONTINUE
C      INTERPOLATION AND PLOT OF PRESSURE AND ROTATIONAL TRANSFORM
       DR=RP(NI)/49.0
       DO 30 I=1,50
    30 SP(I)=(I-1)*DR
       IND=10
       CALL SPLIF (1,NI,RP1,QQ,FP,FPP,FPPP,3,0.,3,0.,0,0.0,IND)
       CALL INTPL (1,50,SP,SI,1,NI,RP1,QQ,FP,FPP,FPPP,0)
       CALL SPLIF (1,NI,RP1,PR,FP,FPP,FPPP,1,0.,3,0.,0,0.,IND)
       CALL INTPL (1,50,SP,SJ,1,NI,RP1,PR,FP,FPP,FPPP,0)
       XD=3.0
       YD=7.0
       TEXTX(1)=8H       RA
       TEXTX(2)=8HDIUS
       TEXTX(3)=8H
       TEXTY(1)=8H     PRES
       TEXTY(2)=8HSURE
       TEXTY(3)=8H
       SYM(1)=8H
       SYM(2)=8H
```

```
          SYM(3)=8H
          SYM(4)=8H
          DO 40 I=1,50
          PL1(I)=SP(I)
       40 PL2(I)=SJ(I)
          NPLO=50
          IF (NVAC.LT.0) GO TO 60
          DR=(RM(NIV)-RM(1))/49.0
          DO 50 I=1,50
          SD(I)=RM(1)+(I-1)*DR
          PL1(I+50)=SD(I)
       50 PL2(I+50)=0.
          NPLO=100
       60 CALL PLOTB (3.,3.,NPLO,1,TEXTX,TEXTY,SYM,XD,YD)
          YD=YD-4.5
          TEXTY(1)=8H        M
          TEXTY(2)=8HU
          TEXTY(3)=8H
          DO 70 I=1,50
       70 PL2(I)=SI(I)
          CALL PLOTB (3.0,3.0,50,1,TEXTX,TEXTY,SYM,XD,YD)
          GO TO 150
C         INTERPOLATION AND PLOT OF MAGNETIC FIELD
       80 CONTINUE
          CALL SPLIF (1,NI,RP1,BTP,FP,FPP,FPPP,3,0.,3,0.,0,0.0,IND)
          CALL INTPL (1,50,SP,SI,1,NI,RP1,BTP,FP,FPP,FPPP,0)
          CALL SPLIF (1,NI,RP1,BPP,FP,FPP,FPPP,3,0.,3,0.,0,0.0,IND)
          CALL INTPL (1,50,SP,SJ,1,NI,RP1,BPP,FP,FPP,FPPP,0)
          XD=XD+5.5
          YD=YD+4.5
          TEXTX(1)=8H        RA
          TEXTX(2)=8HDIUS
          TEXTX(3)=8H
          TEXTY(1)=8H MAGNETI
          TEXTY(2)=8HC FIELD
          TEXTY(3)=8H
          SYM(1)=8HBT
          SYM(2)=8HBP
          DO 90 I=1,50
          PL1(I)=SP(I)
          PL2(I)=SI(I)
       90 PL3(I)=SJ(I)
          NPLO=50
          IF (NVAC.LT.0) GO TO 110
          CALL SPLIF (1,N6,RV1,BTV,FP,FPP,FPPP,3,0.,3,0.,0,0.,IND)
          CALL INTPL (1,50,SD,SIV,1,N6,RV1,BTV,FP,FPP,FPPP,0)
          CALL SPLIF (1,N6,RV1,BPV,FP,FPP,FPPP,3,0.,3,0.,0,0.,IND)
          CALL INTPL (1,50,SD,SJV,1,N6,RV1,BPV,FP,FPP,FPPP,0)
          DO 100 I=1,50
          PL1(I+50)=SD(I)
          PL2(I+50)=SIV(I)
      100 PL3(I+50)=SJV(I)
          NPLO=100
```

```
    110 CALL PLOTB (3.0,3.0,NPLO,2,TEXTX,TEXTY,SYM,XD,YD)
C       INTERPOLATION AND PLOT OF CURRENT
        YD=YD-4.5
        TEXTY(1)=8H        CU
        TEXTY(2)=8HRRENT
        TEXTY(3)=8H
        SYM(1)=8HIT
        SYM(2)=8HIP
        CALL SPLIF (1,NI,RP1,TC,FP,FPP,FPPP,3,0.,3,0.,0,0.0,IND)
        CALL INTPL (1,50,SP,SI,1,NI,RP1,TC,FP,FPP,FPPP,0)
        CALL SPLIF (1,NI,RP1,PC,FP,FPP,FPPP,3,0.,3,0.,0,0.0,IND)
        CALL INTPL (1,50,SP,SJ,1,NI,RP1,PC,FP,FPP,FPPP,0)
        DO 120 I=1,50
        PL2(I)=SI(I)
    120 PL3(I)=SJ(I)
        IF (NVAC.LT.0) GO TO 140
        DO 130 I=1,50
        PL2(I+50)=C1
    130 PL3(I+50)=C2
    140 CALL PLOTB (3.0,3.0,NPLO,2,TEXTX,TEXTY,SYM,XD,YD)
        GO TO 440
    150 CONTINUE
        XD=XD+5.5
        YD=YD+4.5
        N3=NI-1
        N33=NI-2
        DO 160 I=1,N3
    160 D1(I)=SNL2(I)
        DO 170 I=1,50
    170 PL1(I)=D1(3)+(D1(N33)-D1(3))*(I-1)/49.0
        CALL SPLIF (3,N33,D1,AMERN,FP,FPP,FPPP,3,0.0,3,0.0,0,0.0,IND)
        CALL INTPL (1,50,PL1,PL2,3,N33,D1,AMERN,FP,FPP,FPPP,0)
        TEXTX(1)=8H        FLUX
        TEXTX(2)=8H
        TEXTX(3)=8H
        TEXTY(1)=8HMERCIER
        TEXTY(2)=8HCRITERIO
        TEXTY(3)=8HN
        SYM(1)=8H
        NPLO=50
        CALL PLOTB (3.0,3.0,NPLO,1,TEXTX,TEXTY,SYM,XD,YD)
        YD=YD-4.5
        TEXTY(1)=8HISLAND W
        TEXTY(2)=8HIDTH
        TEXTY(3)=8H
        DO 180 I=1,50
    180 PL1(I)=D1(3)+(D1(N33)-D1(3))*(I-1)/49.0
        CALL SPLIF (3,N33,D1,CNORT,FP,FPP,FPPP,3,0.0,3,0.0,0,0.0,IND)
        CALL INTPL (1,50,PL1,PL2,3,N33,D1,CNORT,FP,FPP,FPPP,0)
        CALL PLOTB (3.0,3.0,NPLO,1,TEXTX,TEXTY,SYM,XD,YD)
        GO TO 400
C       PLOT OF CROSS SECTIONS
    190 XOR(1)=2.0
```

```
        XOR(2)=5.7
        XOR(3)=2.0
        XOR(4)=5.7
        YOR(1)=7.5
        YOR(2)=7.5
        YOR(3)=3.5
        YOR(4)=3.5
        DF=1.0/99.0
        IINI=2
        ISTEP=1
        IF (NI.GT.17) ISTEP=2
        IF (NI.GT.17) IINI=3
        DO 200 J=2,N2
200     SP(J)=(J-2)*HU
        DO 210 I=1,100
210     SD(I)=(I-1)*DF
        XY=FLOAT(NK)/4.0
        NN=INT(XY)
        AA=0.0
        DO 220 L=1,4
        K=2+INT(FLOAT(L-1)*XY)
        DO 220 J=2,N1
        CC=ABS(R(J,K))
        DD=ABS(Z(J,K))
220     AA=AMAX1(AA,CC,DD)
        SCA=1.4/AA
        IPL=0
        DO 360 L=1,4
        K=2+INT(FLOAT(L-1)*XY)
        IPL=IPL+1
        IF (IPL.GT.4) GO TO 370
        THE(IPL)=FLOAT(K-2)/FLOAT(NK)
        X1=-1.5+XOR(IPL)+1.0
        X2=1.5+XOR(IPL)+1.0
        Y1=YOR(IPL)+1.0
        Y2=Y1
        CALL LINE (X1,Y1,X2,Y2)
        X1=XOR(IPL)+1.0
        Y1=-1.5+YOR(IPL)+1.0
        Y2=Y1+3.0
        CALL LINE (X1,Y1,X1,Y2)
        DO 270 I=IINI,NI,ISTEP
        DO 230 J=2,N1
        X1=0.5*(RVA(K)+RVA(K+1))
        X2=0.25*(X(J,K)+X(J+1,K)+X(J,K+1)+X(J+1,K+1))
        X3=0.25*(RO(I,J,K)+RO(I,J+1,K)+RO(I,J,K+1)+RO(I,J+1,K+1))
        X3=X3*SL1(I)
        X4=0.5*(RA(K)+RA(K+1))
        X1=X1+X2*(R(J,K)-X1)
230     F(J)=X4+X3*(X1-X4)
        F(1)=F(N1)
        F(N2)=F(2)
        VM=(F(3)+F(1)-2.0*F(2))/(HU*HU)
```

```
      CALL SPLIF (2,N2,SP,F,FP,FPP,FPPP,2,VM,2,VM,0,0.0,IND)
      CALL INTPL (1,100,SD,SI,2,N2,SP,F,FP,FPP,FPPP,0)
      DO 240 J=2,N1
      X1=0.5*(ZVA(K)+ZVA(K+1))
      X2=0.25*(X(J,K)+X(J+1,K)+X(J,K+1)+X(J+1,K+1))
      X3=0.25*(RO(I,J,K)+RO(I,J+1,K)+RO(I,J,K+1)+RO(I,J+1,K+1))
      X3=X3*SL1(I)
      X4=0.5*(ZA(K)+ZA(K+1))
      X1=X1+X2*(Z(J,K)-X1)
  240 F(J)=X4+X3*(X1-X4)
      F(1)=F(N1)
      F(N2)=F(2)
      VM=(F(3)+F(1)-2.0*F(2))/(HU*HU)
      CALL SPLIF (2,N2,SP,F,FP,FPP,FPPP,2,VM,2,VM,0,0.0,IND)
      CALL INTPL (1,100,SD,SJ,2,N2,SP,F,FP,FPP,FPPP,0)
      DO 250 J=1,100
      SI(J)=SI(J)*SCA+XOR(IPL)+1.0
  250 SJ(J)=SJ(J)*SCA+YOR(IPL)+1.0
      CALL SETCRT (SI(1),SJ(1))
      DO 260 J=1,100
  260 CALL VECTOR (SI(J),SJ(J))
      RAXIS=0.5*(RA(K)+RA(K+1))*SCA+XOR(IPL)+1.0
      ZAXIS=0.5*(ZA(K)+ZA(K+1))*SCA+YOR(IPL)+1.0
      CALL SETLCH (RAXIS,ZAXIS,1,0,0,0)
      CALL CRTBCD (1HO)
  270 CONTINUE
C     IF UPLOT.GT.0, RAYS U=CONSTANT ARE PLOTTED ON EACH CROSS SECTION
      UPLOT=1.0
      IF (UPLOT.LT.0.0) GO TO 290
      JSTEP=1
      IF (NJ.GT.24) JSTEP=2
      DO 280 J=2,N1,JSTEP
      X1=0.25*(X(J,K)+X(J,K+1)+X(J+1,K)+X(J+1,K+1))
      X2=0.5*(RVA(K)+RVA(K+1))
      X3=0.5*(ZVA(K)+ZVA(K+1))
      X4=0.5*(RA(K)+RA(K+1))
      X5=0.5*(ZA(K)+ZA(K+1))
      X6=X2+X1*(R(J,K)-X2)
      X7=X3+X1*(Z(J,K)-X3)
      X4=X4*SCA+XOR(IPL)+1.0
      X5=X5*SCA+YOR(IPL)+1.0
      X6=X6*SCA+XOR(IPL)+1.0
      X7=X7*SCA+YOR(IPL)+1.0
      CALL SETCRT (X4,X5)
      CALL VECTOR (X6,X7)
  280 CONTINUE
  290 CONTINUE
      IF (NVAC.LT.0) GO TO 350
      DO 300 J=2,N1
  300 F(J)=R(J,K)
      F(1)=F(N1)
      F(N2)=F(2)
      VM=(F(3)+F(1)-2.0*F(2))/(HU*HU)
```

```
      CALL SPLIF (2,N2,SP,F,FP,FPP,FPPP,2,VM,2,VM,0,0.0,IND)
      CALL INTPL (1,100,SD,SI,2,N2,SP,F,FP,FPP,FPPP,0)
      DO 310 J=2,N1
  310 F(J)=Z(J,K)
      F(1)=F(N1)
      F(N2)=F(2)
      VM=(F(3)+F(1)-2.0*F(2))/(HU*HU)
      CALL SPLIF (2,N2,SP,F,FP,FPP,FPPP,2,VM,2,VM,0,0.0,IND)
      CALL INTPL (1,100,SD,SJ,2,N2,SP,F,FP,FPP,FPPP,0)
      DO 320 J=1,100
      SI(J)=SI(J)*SCA+XOR(IPL)+1.0
  320 SJ(J)=SJ(J)*SCA+YOR(IPL)+1.0
      CALL SETCRT (SI(1),SJ(1))
      DO 330 J=1,100
  330 CALL VECTOR (SI(J),SJ(J))
      GO TO 350
  350 CONTINUE
      XRA=0.5*(RA(K)+RA(K+1))*SCA-0.0
      YRA=0.5*(ZA(K)+ZA(K+1))*SCA-0.0
      XRA=XRA+XOR(IPL)+1.0
      YRA=YRA+YOR(IPL)+1.0
      CALL SETLCH (XRA,YRA,1,0,2,0)
      XRA=-XOR(IPL)
      ZRA=-YOR(IPL)
  360 CONTINUE
  370 CONTINUE
      RINR=RX
      ASP=ZLE/(2.0*PI*RX*NRUN)
      IF (EP.LT.0.00001) GO TO 380
      ENCODE (54,580,T) THE(1),THE(2),THE(3),THE(4),ASP
      CALL SETLCH (1.5,2.25,1,0,1,0)
      CALL CRTBCD (T,7)
      RAYR=1./EP
      ENCODE (54,590,T) RAYR,RINR
      GO TO 390
  380 ENCODE (54,600,T) THE(1),THE(2),THE(3),THE(4),ASP
      CALL SETLCH (1.5,2.25,1,0,1,0)
      CALL CRTBCD (T,7)
      ENCODE (54,610,T) RINR
  390 CALL SETLCH (1.5,1.9,1,0,1,0)
      CALL CRTBCD (T,7)
      CALL FRAME
      GO TO 20
C     ENERGY PLOT
  400 REWIND 3
      REWIND 4
      XY=FLOAT(ITER)/FLOAT(IC)
      N=INT(XY-1.0)
      IF (N.GT.1500) N=1500
      YD=YD+4.5
      XD=XD-5.5
      CALL FRAME
      DO 410 I=1,N
```

```
          PL1(I)=(I-1)*DT*IC
          READ (3) PL2(I),ERO,EAL,EAX
          READ (3) (XR(M),M=1,7),(XZ(M),M=1,7)
          READ (3) ((YB(J,K),K=1,4),J=1,4)
      410 READ (3) EA1,EA2,EA3
          REWIND 3
          TEXTX(1)=8HARTIFICI
          TEXTX(2)=8HAL TIME
          TEXTX(3)=8H
          TEXTY(1)=8H     LOG1
          TEXTY(2)=8HO ENERGY
          TEXTY(3)=8H
          SYM(1)=8H
          PLL=PL2(1)
          DO 420 I=1,N
          X1=ABS(PL2(I)/PLL)
      420 PL2(I)=ALOG10(X1)
          CALL PLOTB (3.0,3.0,N,1,TEXTX,TEXTY,SYM,XD,YD)
          YD=YD-4.5
C         PLOT OF PLASMA RESIDUALS
          DO 430 I=1,N
          READ (3) ETOT,ERO,EAL,EAX
          READ (3) (XR(M),M=1,7),(XZ(M),M=1,7)
          READ (3) ((YB(J,K),K=1,4),J=1,4)
          READ (3) EA1,EA2,EA3
          PL2(I)=ALOG10(ERO)
          PL3(I)=ALOG10(EAL)
          PL4(I)=ALOG10(EAX)
      430 CONTINUE
          REWIND 3
          TEXTY(1)=8HLOG10 RE
          TEXTY(2)=8HSIDUAL P
          TEXTY(3)=8HLASMA
          SYM(1)=8HR
          SYM(2)=8HPSI
          SYM(3)=8HAXIS
          CALL PLOTB (3.0,3.0,N,3,TEXTX,TEXTY,SYM,XD,YD)
          GO TO 520
      440 CONTINUE
          IF(NVAC.LT. 0) GO TO 550
          REWIND 4
          YD=YD+4.5
          XD=XD-5.5
          CALL FRAME
C         PLOT OF VACUUM RESIDUALS
          DO 450 I=1,N
          PL1(I)=(I-1)*DT*IC
          READ (4) ERBO,ERBO1,ERVA
          READ (4) (YR(M),M=1,7),(YZ(M),M=1,7)
          READ (4) OM1,OM2
          PL2(I)=ALOG10(ERBO)
          PL3(I)=ALOG10(ERBO1)
          PL4(I)=ALOG10(ERVA)
```

```
      450 CONTINUE
          REWIND 4
          TEXTX(1)=8HARTIFICI
          TEXTX(2)=8HAL TIME
          TEXTX(3)=8H
          TEXTY(1)=8HLOG10 RE
          TEXTY(2)=8HSIDUAL V
          TEXTY(3)=8HACUUM
          SYM(1)=8HBOUNDARY
          SYM(2)=8HAXIS
          SYM(3)=8HVACUUM
          CALL PLOTB (3.,3.,N,3,TEXTX,TEXTY,SYM,XD,YD)
          YD=YD-4.5
C         PLOT OF FOURIER COEFFICIENTS FOR THE VACUUM AXIS
          DO 460 I=1,N
          READ (4) ERBO,ERBO1,ERVA
          READ (4) (YR(M),M=1,7),(YZ(M),M=1,7)
          READ (4) OM1,OM2
          PL2(I)=YR(NRA1)
          PL3(I)=YR(NRA2)
          PL4(I)=YZ(NZA1)
          PL5(I)=YZ(NZA2)
      460 CONTINUE
          REWIND 4
          TEXTY(1)=8HFOURIER
          TEXTY(2)=8HCOEFF VA
          TEXTY(3)=8HC AXIS
          SYM(1)=RNAME(NRA1)
          SYM(2)=RNAME(NRA2)
          SYM(3)=ZNAME(NZA1)
          SYM(4)=ZNAME(NZA2)
          CALL PLOTB (3.,3.,N,4,TEXTX,TEXTY,SYM,XD,YD)
          GO TO 550
      470 CONTINUE
          REWIND 3
          YD=YD+4.5
          XD=XD-5.5
          CALL FRAME
C         PLOT OF FOURIER COEFFICIENTS FOR THE MAGNETIC AXIS
          DO 480 I=1,N
          READ (3) ETOT,ERO,EAL,EAX
          READ (3) (XR(M),M=1,7),(XZ(M),M=1,7)
          READ (3) ((YB(J,K),K=1,4),J=1,4)
          READ (3) EA1,EA2,EA3
          PL2(I)=XR(NRA1)
          PL3(I)=XR(NRA2)
          PL4(I)=XZ(NZA1)
          PL5(I)=XZ(NZA2)
      480 CONTINUE
          REWIND 3
          TEXTY(1)=8HFOURIER
          TEXTY(2)=8HCOEFF MA
          TEXTY(3)=8HG AXIS
```

```
      SYM(1)=RNAME(NRA1)
      SYM(2)=RNAME(NRA2)
      SYM(3)=ZNAME(NZA1)
      SYM(4)=ZNAME(NZA2)
      CALL PLOTB (3.,3.,N,4,TEXTX,TEXTY,SYM,XD,YD)
      YD=YD-4.5
C     PLOT OF FOURIER COEFFICIENTS FOR A FLUX SURFACE
      DO 490 I=1,N
      READ (3) ETOT,ERO,EAL,EAX
      READ (3) (XR(M),M=1,7),(XZ(M),M=1,7)
      READ (3) ((YB(J,K),J=1,4),K=1,4)
      READ (3) EA1,EA2,EA3
      PL2(I)=YB(M1+1,K1+1)
      PL3(I)=YB(M2+1,K2+1)
      PL4(I)=YB(M3+1,K3+1)
      PL5(I)=YB(M4+1,K4+1)
  490 CONTINUE
      REWIND 3
      IF (NVAC.LT.0) GO TO 500
      TEXTY(1)=8HFOURIER
      TEXTY(2)=8HCOEFF FR
      TEXTY(3)=8HEE BOUND
      GO TO 510
  500 TEXTY(1)=8HFOURIER
      TEXTY(2)=8HCOEFF RA
      TEXTY(3)=8HDIUS
  510 TSYM=3HMK=
      ENCODE (10,620,SYM(1)) TSYM,M1,K1
      ENCODE (10,620,SYM(2)) TSYM,M2,K2
      ENCODE (10,620,SYM(3)) TSYM,M3,K3
      ENCODE (10,620,SYM(4)) TSYM,M4,K4
      CALL PLOTB (3.,3.,N,4,TEXTX,TEXTY,SYM,XD,YD)
      GO TO 80
C     PLOT OF DESCENT COEFFICIENTS
  520 CONTINUE
      REWIND 3
      XD=XD+5.5
      YD=YD+4.5
      DO 530 I=1,N
      READ (3) ETOT,ERO,EAL,EAX
      READ (3) (XR(M),M=1,7),(XZ(M),M=1,7)
      READ (3) ((YB(J,K),K=1,4),J=1,4)
      READ (3) PL2(I),PL3(I),PL4(I)
      PL2(I)=ALOG10(PL2(I))
  530 CONTINUE
      TEXTY(1)=8HLOG10 DE
      TEXTY(2)=8HSCENT CO
      TEXTY(3)=8HEFF
      SYM(1)=8HE1/A1
      CALL PLOTB (3.,3.,N,1,TEXTX,TEXTY,SYM,XD,YD)
      YD=YD-4.5
      DO 540 I=1,N
      PL2(I)=ALOG10(PL3(I))
```

```
      540 PL3(I)=ALOG10(PL4(I))
          TEXTY(1)=8HLOG10 RE
          TEXTY(2)=8HSIDUAL P
          TEXTY(3)=8HLASMA
          SYM(1)=8HR AXIS
          SYM(2)=8HMU
          CALL PLOTB (3.,3.,N,2,TEXTX,TEXTY,SYM,XD,YD)
          GO TO 470
      550 CONTINUE
          RETURN
C
      570 FORMAT (25HI.D. NUMBER 110801   RUN=I2)
      580 FORMAT (21HCROSS SECTIONS AT V= F3.2,1H,F3.2,1H,F3.2,1H,F3.2,13H,
         11/(EP*QLZ)=F5.2)
      590 FORMAT (13HMAJOR RADIUS=F6.2,17X,13HMINOR RADIUS=F5.2)
      600 FORMAT (21HCROSS SECTIONS AT V= F3.2,1H,F3.2,1H,F3.2,1H,F3.2,11H,
         1QLZ/2*PI=F5.2)
      610 FORMAT (21HMAJOR RADIUS INFINITE,12X,14H MINOR RADIUS=F5.2)
      620 FORMAT (A3,2I1)
          END




          SUBROUTINE PLOTB (XL,YL,N,M,TEXTX,TEXTY,SYM,XD,YD)
C         PLOTS XPL(I,J) FOR 1.LT.J AND J.LE.5 AS A FUNCTION OF XPL(I,1).
C         XL AND YL ARE THE LENGTHS IN INCHES OF THE X AND Y AXES. N IS THE
C         NUMBER OF POINTS PLOTTED FOR EACH CURVE, M IS THE NUMBER OF
C         CURVES. TEXTX AND TEXTY ARE THE LABELS FOR THE X AND Y AXES.
C         SYM(J) IS THE LABEL FOR THE J-TH CURVE.
          USE NAME17
          DIMENSION TEXTX(3), TEXTY(3), XA(2), XB(2), SYM(4)
          XMIN=1000.
          XMAX=-1000.
          XL=5./6.*XL
          YL=5./6.*YL
          MM=M+1
          YMAX=-1000.
          YMIN=1000.
          DO 10 I=1,N
          XMAX=AMAX1(XMAX,XPL(I,1))
          XMIN=AMIN1(XMIN,XPL(I,1))
          DO 10 J=2,MM
          YMAX=AMAX1(YMAX,XPL(I,J))
       10 YMIN=AMIN1(YMIN,XPL(I,J))
          IF (XMAX-XMIN.LE.0.001) XMAX=XMIN+0.001
          IF (YMAX-YMIN.LE.0.001) YMAX=YMIN+0.001
          XA(1)=XMIN
          XA(2)=XMAX
```

```
      XB(1)=YMIN
      XB(2)=YMAX
      CALL GAXIS (XD,YD,XD+XL,YD,0,0,0,"F7.2",3,XA)
      CALL GAXIS (XD,YD,XD,YD+YL,0,0,1,"E8.1",3,XB)
      CALL SETLCH (XD+0.5,YD-0.5,1,0,1,0)
      CALL CRTBCD (TEXTX,3)
      CALL SETLCH (XD-1.2,YD+0.25,1,0,1,1)
      CALL CRTBCD (TEXTY,3)
      XO=XMIN
      YO=YMIN
      DY=(YMAX-YMIN)/YL
      DX=(XMAX-XMIN)/XL
      DO 20 I=1,N
      DO 20 J=2,MM
   20 XPL(I,J)=(XPL(I,J)-YO)/DY+YD
      DO 40 J=2,MM
      XPL1=(XPL(1,1)-XO)/DX+XD
      CALL SETCRT (XPL1,XPL(1,J))
      DO 30 I=1,N
      XPL1=(XPL(I,1)-XO)/DX+XD
   30 CALL VECTOR (XPL1,XPL(I,J))
      CALL SETLCH (XPL1+.1,XPL(N,J),1,0,1,0)
      CALL CRTBCD (SYM(J-1))
   40 CONTINUE
      RETURN
      END




      SUBROUTINE PBOU
C     COMPUTES BOUNDARY VALUES FOR VACUUM POTENTIAL. THIS SUBROUTINE
C     SHOULD BE CHANGED IF A MORE GENERAL WINDING LAW IS DESIRED.
      USE NAME1
      USE NAME2
      USE NAME6
      USE NAME9
      USE NAME10
      USE NAME12
      USE NAME13
      USE NAME14
      USE NAME15
      DIMENSION AN(5)
      PI2=2.0*PI
      FAXV=VERT
      XR=1.0/WRAD
      DO 20 J=2,N1
      DO 10 K=2,N4
      G2(J,K)=0.25*(R(J,K)+R(J-1,K)+R(J,K-1)+R(J-1,K-1))
```

```
   10 AKB(J,K)=0.25*(Z(J,K)+Z(J-1,K)+Z(J,K-1)+Z(J-1,K-1))
      AKB(J,1)=AKB(J,N4)
      G2(J,1)=G2(J,N4)
      G2(J,N5)=G2(J,2)
   20 AKB(J,N5)=AKB(J,2)
      DO 30 K=1,N5
      AKB(1,K)=AKB(N1,K)
      G2(1,K)=G2(N1,K)
      G2(N2,K)=G2(2,K)
   30 AKB(N2,K)=AKB(2,K)
      DO 70 J=1,N2
      U=(J-2.5)*HU
      UP1=PI2*U
      UP2=UP1*AK3
      Y1=KW1*UP1+TORS*SIN(UP2)
      DO 70 K=1,N5
      V=(K-2.5)*HV
      VP1=PI2*V
      Y2=C1*U+VERT*AKB(J,K)
      YY=Y1-KW2*VP1-US0*PI2
      RAD=WRAD
      X1=XR*SIN(YY)/(1.0-XR*COS(YY))
      SUM=ATAN(X1)/XR
   70 PT(NIV,J,K)=Y2+C2*V+SUM*AMPH
      RETURN
C
   80 FORMAT (3X,"L=",I2,3X,"V=",F6.3,3X,"U=",F6.3,3X,2F10.3)
      END




      SUBROUTINE ASIN
C     EVALUATES FLUX AND MASS FUNCTIONS AND INITIALIZES AXIALLY
C     SYMMETRIC SOLUTION. THIS SUBROUTINE SHOULD BE CHANGED IF MORE
C     GENERAL PRESSURE AND ROTATIONAL TRANSFORM PROFILES OR AXIALLY
C     SYMMETRIC WALL SHAPE ARE DESIRED.
      USE NAME3
      USE NAME7
      USE NAME8
      USE NAME10
      USE NAME12
      USE NAME13
      USE NAME14
      ALFU1=0.0
      IF (NAS.GT.0) ALFU1=ALFU
      DO 40 J=1,N2
      U=(J-2)*HU
      U1=U
```

```
        UP1=2.0*PI*U1
        UBAR=UP1+ALFU1*SIN(UP1)
        UBARP=2.0*PI*(1.0+ALFU1*COS(UP1))
        X1=COS(UBAR)
        X2=SIN(UBAR)
        X3=COS(2.0*UBAR)
        X4=SIN(2.0*UBAR)
        X5=COS(3.0*UBAR)
        X6=SIN(3.0*UBAR)
C       COMPUTE WALL SHAPE
        IF (NAS.LT.0) GO TO 10
        RAD=1.0+DEL10*X1+DEL20*X3+DEL30*X5
        RADU=-UBARP*(DEL10*X2+2.0*DEL20*X4+3.0*DEL30*X6)
        GO TO 20
     10 RAD=1.0
        RADU=0.0
     20 RR(J)=RAD*X1
        ZZ(J)=RAD*X2
        RU(J)=RADU*X1-UBARP*X2*RAD
        ZU(J)=RADU*X2+UBARP*X1*RAD
C       COMPUTE FREE BOUNDARY FUNCTION
        IF (NVAC.LT.0) GO TO 30
        X(J)=RBOU
        GO TO 40
     30 X(J)=1.0
     40 CONTINUE
        RA=0.0
        ZA=0.0
        R2=0.0
        Z2=0.0
        GUM=1.0/GAM
        FQ=PI*RBOU*RBOU
        AA=AMAX1(ABS(AMU0),ABS(AMU1),ABS(AMU2))
        ALF=0.5*(1.0+ALF)
        SL1(1)=0.0
        SL2(1)=0.0
        XS(1)=0.0
        DO 50 I=2,NI
        S=(I-1)*HS
        SL1(I)=S**ALF
        SL2(I)=SL1(I)*SL1(I)
     50 XS(I)=SL2(I)/HS
        DO 60 I=1,N3
        SA(I)=(SL2(I+1)-SL2(I))/HS
        PS(I)=SL2(I)/(SL2(I+1)-SL2(I))
        QS(I)=SL2(I+1)/(SL2(I+1)-SL2(I))
        DP(I)=SA(I)*(1.0-PS(I))
     60 DQ(I)=SA(I)*(1.0+QS(I))
        PQS(1)=0.5*SA(1)
        PQS(NI)=0.5*SA(N3)
        DO 70 I=2,N3
     70 PQS(I)=0.5*(SA(I)+SA(I-1))
        DO 100 I=1,NI
```

```
      S=(I-1)*HS
      X2=SL1(I)
      X1=SL2(I)
      OP1(I)=FQ
C     COMPUTE ROTATIONAL TRANSFORM PER UNIT LENGTH
      AMU=(AMU0+AMU1*X2+AMU2*X1)/ZLE
      PRES=P0*((1.0-ZPR*X1**YPR)**XPR)
C     COMPUTE FLUX FUNCTIONS
      IF (AA.GT.0.30) GO TO 80
C     IF AMU IS SMALL, TOROIDAL FLUX FUNCTION IS CHOSEN TO SATISFY
C     0.5*BT*BT+P=CONST
      QT(I)=-FQ*SQRT(1.0-2.0*PRES)
      GO TO 90
C     IF AMU IS LARGE, TOROIDAL FLUX FUNCTION IS CHOSEN TO SATISFY BT=1
   80 QT(I)=-FQ
C     POLOIDAL FLUX FUNCTION IS CHOSEN TO SATISFY Q/QT=AMU
   90 Q(I)=AMU*QT(I)
      QQ(I)=Q(I)*Q(I)*SL2(I)
      AM(I)=(PRES**GUM)*FQ
C     COMPUTE INITIAL VALUES OF R AND PSI
      DO 100 J=1,N2
      R(I,J)=1.0
      U=(J-2.5)*HU
  100 AL(I,J)=-U*QT(I)
      DO 101 I=2,N3
      X1=(SL2(I)**(YPR-1.0))/(1.0-ZPR*(SL2(I)**YPR))
      AMS(I)=-XPR*YPR*ZPR*ZLE*GUM*X1*AM(I)
  101 CONTINUE
      SNL2(1)=0.0
      DO 110 I=2,NI
      X1=ABS(0.5*(QT(I)+QT(I-1)))
  110 SNL2(I)=SNL2(I-1)+X1*(SL2(I)-SL2(I-1))/(PI*RBOU*RBOU)
C     COMPUTE MASS FUNCTION FROM GIVEN INITIAL PRESSURE DISTRIBUTION
      IF (NVAC.LT.0) GO TO 130
C     COMPUTE INITIAL VALUES OF VACUUM POTENTIAL
      DO 120 J=1,N2
      U=(J-2)*HU
      DO 120 I=1,NIV
  120 PA(I,J)=U*C1
  130 CONTINUE
      RETURN
      END
```

```
      SUBROUTINE SURF
C     COMPUTES OUTER WALL SHAPE AND INITIALIZES 3D SOLUTION.
C     THIS SUBROUTINE SHOULD BE CHANGED IF A MORE GENERAL 3D WALL
C     SHAPE IS DESIRED. ALSO COMPUTES TEST FUNCTION FOR STABILITY.
      USE NAME1
      USE NAME2
      USE NAME6
      USE NAME9
      USE NAME10
      USE NAME12
      USE NAME13
      USE NAME14
      USE NAME16
      USE NAME20
      DIMENSION AA(5,5), BB(5), CC(5), AW(5,5), WKS1(5), WKS2(5)
      PI2=2.0*PI
      DO 10 I=1,NI
   10 AM(I)=AM(I)*ZLE
      DO 20 I=1,NI
   20 Q(I)=Q(I)*ZLE
      SUM=0.0
      DO 190 K=1,N5
      V=(K-2.5)*HV
      VV=2.0*PI*V
      X2=SIN(VV)
      X3=COS(VV)
      X4=SIN(2.0*VV)
      X5=COS(2.0*VV)
      X6=SIN(3.0*VV)
      X7=COS(3.0*VV)
C     COMPUTE INITIAL VALUES OF MAGNETIC AND VACUUM AXES
      X8=1.0-X3
      X9=DELA
      IF (NAS.LT.0) GO TO 30
      RA(K)=RRA
      ZA(K)=ZZA
      RVA(K)=RR2
      ZVA(K)=ZZ2
      GO TO 70
   30 CONTINUE
      IF (NGEOM.LT.3) GO TO 50
      IF (NGEOM.GT.3) GO TO 40
      RA(K)=(DELB+X9)*X3
      ZA(K)=-(DELB+X9)*X2
      GO TO 60
   40 CONTINUE
      RA(K)=RRA+DEL1R*X3
      ZA(K)=ZZA+DEL1Z*X2
      GO TO 60
   50 CONTINUE
      RA(K)=RRA+DEL1*X3
      ZA(K)=ZZA+DEL1*X2
   60 CONTINUE
```

```
      RVA(K)=RA(K)
      ZVA(K)=ZA(K)
   70 CONTINUE
      VP1=(V+0.5*HV)*PI2
      VP2=2.0*VP1
      VP3=3.0*VP1
      X1=COS(VP1)
      X2=SIN(VP1)
      DO 190 J=1,N2
      U=(J-2.5)*HU
      UP1=(U+0.5*HU)*PI2
      UP2=2.0*UP1
      UP3=3.0*UP1
      UBAR=UP1+ALFU*SIN(UP1)
      UBARP=PI2*(1.0+ALFU*COS(UP1))
      UP2=2.0*UBAR
      UP3=3.0*UBAR
      Y1=COS(UBAR)
      Y2=SIN(UBAR)
      Y3=SIN(UP2-VP1)
      Y4=SIN(UP3-VP1)
      Y5=SIN(UP3-VP3)
      Y6=SIN(UP2-VP2)
C     COMPUTE WALL SHAPE AND DERIVATIVES OF WALL FUNCTION
      GO TO (110,120,80,90) ,NGEOM
   80 CONTINUE
      TT1=-VP1
      TT1V=-PI2
      T1=TT1+DELC*SIN(UBAR)
      T1V=TT1V
      T2=1.0-X1
      RBAR=DELA
      RBARV=0.0
      RAD=DELB*(1.0+DEL2*COS(UBAR))
      RADU=-DELB*DEL2*SIN(UBAR)
      RADV=0.0
      T1U=DELC*COS(UBAR)
      R(J,K)=RBAR*COS(TT1)+RAD*COS(T1)
      Z(J,K)=RAD*SIN(T1)+RBAR*SIN(TT1)
      RU(J,K)=(RADU*COS(T1)-RAD*SIN(T1)*T1U)*UBARP
      ZU(J,K)=(RADU*SIN(T1)+RAD*COS(T1)*T1U)*UBARP
      RV(J,K)=RBARV*COS(TT1)+RADV*COS(T1)-RAD*SIN(T1)*T1V-RBAR*SIN(TT1)
     1 *TT1V
      ZV(J,K)=RADV*SIN(T1)+RAD*COS(T1)*T1V+RBARV*SIN(TT1)+RBAR*COS(TT1)
     1 *TT1V
      GO TO 140
   90 CONTINUE
      SUCO=0.0
      SUSI=0.0
      SUCOU=0.0
      SUSIU=0.0
      SUCOV=0.0
      SUSIV=0.0
```

```
      DO 100 L=1,3
      DO 100 M=1,4
      XX1=COS(L*UP1-(M-1)*VP1)*DEL(L,M)
      XX2=SIN(L*UP1-(M-1)*VP1)*DEL(L,M)
      SUCO=SUCO+XX1
      SUSI=SUSI+XX2
      SUCOU=SUCOU+L*XX1
      SUSIU=SUSIU+L*XX2
      SUCOV=SUCOV+(M-1)*XX1
      SUSIV=SUSIV+(M-1)*XX2
  100 CONTINUE
      R(J,K)=Y1+DEL1R*X1-SUCO
      Z(J,K)=Y2+DEL1Z*X2+SUSI
      RU(J,K)=PI2*(-Y2+SUSIU)
      ZU(J,K)=PI2*(Y1+SUCOU)
      RV(J,K)=PI2*(-DEL1R*X2-SUSIV)
      ZV(J,K)=PI2*(DEL1Z*X1-SUCOV)
      GO TO 140
  110 CONTINUE
      RAD=1.-DEL0*X1+DEL10*Y1+DEL20*COS(UP2)+DEL30*COS(UP3)-DEL3*COS(UP3
     1 -VP1)+DEL22*COS(UP2-VP2)+DEL33*COS(UP3-VP3)
      RADU=-UBARP*(DEL10*Y2+2.*DEL20*SIN(UP2)+3.*DEL30*SIN(UP3)-3.*DEL3
     1 *Y4+2.0*DEL22*Y6+3.*DEL33*Y5)
      RADV=PI2*(DEL0*X2-DEL3*Y4+2.*DEL22*Y6+3.*DEL33*Y5)
      GO TO 130
  120 CONTINUE
      AN=3.0
      RAD1=1.0+AN*DEL3*COS(3.0*UBAR-VP1)
      RAD=RAD1**(-1.0/AN)
      RAD1=RAD/RAD1
      RADU=RAD1*UBARP*DEL3*3.0*SIN(3.0*UBAR-VP1)
      RADV=-RAD1*PI2*DEL3*SIN(3.0*UBAR-VP1)
  130 CONTINUE
      R(J,K)=RAD*Y1+DEL1*X1-DEL2*COS(UBAR-VP1)
      Z(J,K)=RAD*Y2+DEL1*X2+DEL2*SIN(UBAR-VP1)
      RU(J,K)=RADU*Y1-UBARP*RAD*Y2+UBARP*DEL2*SIN(UBAR-VP1)
      ZU(J,K)=RADU*Y2+UBARP*RAD*Y1+UBARP*DEL2*COS(UBAR-VP1)
      RV(J,K)=RADV*Y1-PI2*DEL1*X2-PI2*DEL2*SIN(UBAR-VP1)
      ZV(J,K)=RADV*Y2+PI2*DEL1*X1-PI2*DEL2*COS(UBAR-VP1)
  140 CONTINUE
      IF (NVAC.LT.0) GO TO 160
      X(J,K)=XX(J)
      DO 150 I=1,NI
  150 RO(I,J,K)=E1(I,J)
      GO TO 180
  160 X(J,K)=1.0
      DO 170 I=1,NI
      S=(I-1)*HS
  170 RO(I,J,K)=E1(I,J)
  180 CONTINUE
C     COMPUTE INITIAL VALUES FOR PSI
      DO 190 I=1,NI
      S=(I-1)*HS
```

```
            AL(I,J,K)=E2(I,J)+Q(I)*V
      190 CONTINUE
            HMAX=-1000.0
            HMIN=1000.0
            DO 200 K=2,N4
            DO 200 J=2,N1
            X1=R(J,K)-0.5*(RA(K)+RA(K+1))
            X2=Z(J,K)-0.5*(ZA(K)+ZA(K+1))
            X3=X1*ZU(J,K)-X2*RU(J,K)
            HMAX=AMAX1(HMAX,X3)
            HMIN=AMIN1(HMIN,X3)
      200 CONTINUE
            PRINT 410, HMAX,HMIN
            IF (NVAC.LT.0) GO TO 220
C         COMPUTE INITIAL VALUES FOR VACUUM POTENTIAL
            CALL PBOU
            DO 210 I=1,N6
            S1=(I-1)*HR
            DO 210 J=1,N2
            U=(J-2.5)*HU
            DO 210 K=1,N5
            V=(K-2.5)*HV
      210 PT(I,J,K)=C1*U+C2*V+(PT(NIV,J,K)-C1*U-C2*V)*S1
      220 CONTINUE
C         COMPUTE TEST FUNCTION FOR STABILITY CALCULATION
            NP5=NK*NRUN+2
            DO 230 K=1,NP5
            V=PI2*(K-2.5)*HV/NRUN
            EH1(K)=COS(V)
            EH2(K)=COS(2.0*V)
            EJ1(K)=SIN(V)
            EJ2(K)=SIN(2.0*V)
            DO 230 J=1,N2
            U=PI2*(J-2.5)*HU
            EF1(J,K)=COS(U-1.0*V)
            EF2(J,K)=COS(2.0*U-V)
            EF3(J,K)=COS(3.0*U-1.0*V)
            EG1(J,K)=SIN(U-1.0*V)
            EG2(J,K)=SIN(2.0*U-V)
      230 EG3(J,K)=SIN(3.0*U-1.0*V)
            IF (NVAC.LT.0) GO TO 250
            DO 240 J=1,N2
            U=PI2*(J-2.5)*HU
            DO 240 K=1,NP5
            V=PI2*(K-2.5)*HV/NRUN
      240 EBD(J,K)=COS(U-V)
      250 CONTINUE
            DO 260 I=1,NI
            S=(I-1)*HS
            EAM1(I)=EAMC1*(1.0-SL2(I))
            EAM2(I)=EAMC2*(1.0-SL2(I))
            EAM3(I)=EAMC3*SL1(I)*(1.0-SL2(I))
            EAM4(I)=EAMC4*(1.0-3.0*SL2(I))
```

```
          EAM5(I)=EAMC5*(1.0-2.0*SL2(I))
  260 EAM6(I)=EAMC6*SL1(I)*(1.0-2.0*SL2(I))
          IF (NAS.GT.0) GO TO 380
C         COMPUTE INITIAL VALUES FOR RO(I,J,K)
          DO 290 K=2,N4
          DO 270 J=1,3
          BB(J)=0.0
          DO 270 JJ=1,3
  270 AA(J,JJ)=0.0
          DO 280 J=2,N1
          X4=0.25*(R(J,K)+R(J,K-1)+R(J-1,K)+R(J-1,K-1))-RA(K)
          X5=0.25*(Z(J,K)+Z(J-1,K)+Z(J,K-1)+Z(J-1,K-1))-ZA(K)
          X4=X4*RBOU
          X5=X5*RBOU
          X1=X4*X4+X5*X5
          Y1=1.0/(RO(1,J,K)*RO(1,J,K))
          XX1=X4*X4
          XX2=2.0*X4*X5
          XX3=X5*X5
          BB(1)=BB(1)+Y1*XX1
          BB(2)=BB(2)+Y1*XX2
          BB(3)=BB(3)+Y1*XX3
          AA(1,1)=AA(1,1)+XX1*XX1
          AA(1,2)=AA(1,2)+XX1*XX2
          AA(1,3)=AA(1,3)+XX1*XX3
          AA(2,2)=AA(2,2)+XX2*XX2
          AA(3,3)=AA(3,3)+XX3*XX3
  280 AA(2,3)=AA(2,3)+XX3*XX2
          AA(2,1)=AA(1,2)
          AA(3,1)=AA(1,3)
          AA(3,2)=AA(2,3)
          CALL FO4ATF (AA,5,BB,3,CC,AW,5,WKS1,WKS2,IFAIL)
          ROAX(K)=CC(1)
          ROBX(K)=CC(2)
          ROCX(K)=CC(3)
  290 CONTINUE
          CALL CBO (N1)
          IF (NAS.GT.0) GO TO 400
          DO 340 J=2,N1
          DO 300 K=1,N4
          RB2(K)=RB1(K)
  300 ZB2(K)=ZB1(K)
          CALL CBO (J)
          DO 310 K=1,N4
          D1(K)=RB1(K)-0.5*(RA(K)+RA(K+1))
          D2(K)=ZB1(K)-0.5*(ZA(K)+ZA(K+1))
          D3(K)=RB2(K)-0.5*(RA(K)+RA(K+1))
          D4(K)=ZB2(K)-0.5*(ZA(K)+ZA(K+1))
          CAF1(K)=D1(K)*D1(K)
          CAG1(K)=2.0*D1(K)*D2(K)
          CAH1(K)=D2(K)*D2(K)
          CAF2(K)=D3(K)*D3(K)
          CAG2(K)=2.0*D3(K)*D4(K)
```

```
          CAH2(K)=D4(K)*D4(K)
310  CONTINUE
          DO 330 K=2,N4
          X1=0.25*(CAF1(K)+CAF1(K-1)+CAF2(K)+CAF2(K-1))
          X2=0.25*(CAG1(K)+CAG1(K-1)+CAG2(K)+CAG2(K-1))
          X3=0.25*(CAH1(K)+CAH1(K-1)+CAH2(K)+CAH2(K-1))
          Y1=ROAX(K)*X1+ROBX(K)*X2+ROCX(K)*X3
          RO(1,J,K)=1.0/SQRT(Y1)
          DO 320 I=2,NI
320  RO(I,J,K)=RO(1,J,K)+SL1(I)*(1.0-RO(1,J,K))
330  CONTINUE
340  CONTINUE
          DO 370 I=1,NI
          DO 350 J=2,N1
          RO(I,J,1)=RO(I,J,N4)
350  RO(I,J,N5)=RO(I,J,2)
          DO 360 K=1,N5
          RO(I,1,K)=RO(I,N1,K)
360  RO(I,N2,K)=RO(I,2,K)
370  CONTINUE
          GO TO 400
380  CONTINUE
          DO 390 K=1,N5
          ROAX(K)=ROA
          ROBX(K)=ROB
          ROCX(K)=ROC
390  CONTINUE
400  CONTINUE
          RETURN
C
410  FORMAT (//,10X,"MAX JAC=",F7.2,3X,"MIN JAC=",F7.2)
          END
```

Index